Ajay Kumar Mishra, Chaudhery Mustansar Hussain and Shivani Bhardwaj Mishra (Eds.)

Nanomaterials for Water Remediation

Also of interest

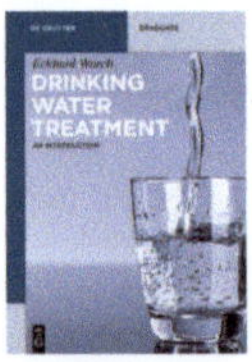

Drinking Water Treatment – An Introduction
Eckhard Worch, 2019
ISBN 978-3-11-055154-9, e-ISBN 978-3-11-055155-6

Aquatic Chemistry – For Water and Wastewater Treatment Applications
Lahav, Birnhack, 2019
ISBN 978-3-11-060392-7, e-ISBN 978-3-11-060395-8

Environmental Functional Nanomaterials
Wang, Zhong (Eds.), 2019
ISBN 978-3-11-054405-3, e-ISBN 978-3-11-054436-7

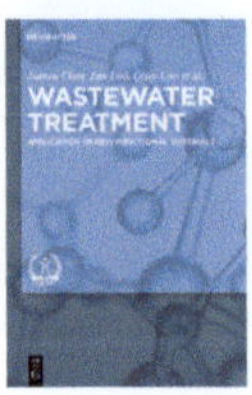

Wastewater Treatment – Application of New Functional Materials
Chen, Luo, Luo, Pang, 2018
ISBN 978-3-11-054278-3, e-ISBN 978-3-11-054438-1

Nanomaterials for Water Remediation

Edited by
Ajay Kumar Mishra, Chaudhery Mustansar Hussain
and Shivani Bhardwaj Mishra

2nd Edition

DE GRUYTER

Editors
Prof. Ajay Kumar Mishra
University of South Africa
Florida Science Campus
JOHANNESBURG 1709
South Africa
ajaykmishra1@gmail.com

Prof. Chaudhery Mustansar Hussain
New Jersey Institute of Technology
323 Martin Luther KIng Jr.Blvd
NEWARK NJ 07103
United States of America
chaudhery.m.hussain@njit.edu

Prof. Shivani Bhardwaj Mishra
University of South Africa
Florida Science Campus
JOHANNESBURG 1709
South Africa
bshivani73@gmail.com

ISBN 978-3-11-064336-7
e-ISBN (PDF) 978-3-11-065060-0
e-ISBN (EPUB) 978-3-11-063455-6

Library of Congress Control Number: 2019958025

Bibliographic information published by the Deutsche Nationalbibliothek
The Deutsche Nationalbibliothek lists this publication in the Deutsche Nationalbibliografie; detailed bibliographic data are available on the Internet at http://dnb.dnb.de.

Cover image: DavidOrr/E+/Getty Images
Typesetting: Integra Software Services Pvt. Ltd.
Printing and binding: CPI books GmbH, Leck

www.degruyter.com

Preface

Nanomaterials are being used to develop more cost-effective and high-performance water treatment systems. Nanomaterials in water research have been extensively utilized for treatment, remediation, and pollution prevention. Remediation is the process of removing toxic pollutants from water. This book entitled *Nanomaterials for Water Remediation* focuses on the carbon-based materials, nanoadsorbent metals, nanoparticles, cryogels, and bentonites for the remediation of various organic and inorganic pollutants from wastewater. Water pollution is mainly caused by pollutants, which lead to severe environmental and health problems. It is a well-established fact that carbon-based materials are very effective for the removal of both organic and inorganic pollutants from wastewater, and nanomaterials have better adsorption capacity, selectivity, and stability than the nanoparticles.

This book broadly covers the fundamental knowledge and recent advancements for the research and development in the field of nanotechnology, environmental science, and water research, which will be highly beneficial to graduate and postgraduate students. The book also provides a platform for all researchers as it covers a huge background for the recent literature and abbreviations.

Ajay Kumar Mishra, Chaudhery Mustansar Hussain, and Shivani Bhardwaj Mishra
Editors

https://doi.org/10.1515/9783110650600-202

Contents

List of contributors

Nityananda Agasti
Department of Chemistry
Deen Dayal Upadhyaya College
University of Delhi
New Delhi, India

Amin Fathi Amin Ajlouni
Department of Chemistry
Taibah University
Madina, Yanbu, Saudi Arabia

Pankaj Attri
Plasma Bioscience Research Center/
Department of Electrical and Biological Physics
Kwangwoon University
Republic of Korea

Aning Ayucitra
Department of Chemical Engineering
National Taiwan University of Science and Technology
Taiwan

Sema Bektaş
Hacettepe University
Department of Chemistry
Ankara, Turkey

A. H. Bhat
Department of Fundamental and Applied Sciences
Universiti Teknologi Petronas Malaysia
Malaysia

Rohit Bhatia
Department of Chemistry, Institute of Home Economics
Delhi, India

Y. K. Dasan
Department of Fundamental and Applied Sciences
Universiti Teknologi Petronas Malaysia
Malaysia

Adil Denizli
Hacettepe University
Department of Chemistry
Ankara, Turkey

Seema Garg
Department of Chemistry
University of Delhi
Delhi, India

Suryadi Ismadji
Department of Chemical Engineering
Widya Mandala Catholic University Surabaya
Surabaya, Indonesia

Yi-Hsu Ju
Department of Chemical Engineering
National Taiwan University of Science and Technology
Taiwan

Farheen Khan
Department of Chemistry
Taibah University
Madina, Yanbu, Saudi Arabia

Imran Khan
CICECO-Aveiro Institute of Materials
Department of Chemistry
University of Aveiro
Portugal

Kumud Malika Tripathi
Department of Material Science & Engineering
Gachon University
Seongam-si, South Korea

Ayşenur Sağlam
Hacettepe University
Department of Chemistry
Ankara, Turkey

https://doi.org/10.1515/9783110650600-204

Felycia Edi Soetaredjo
Department of Chemical Engineering
Widya Mandala Catholic University Surabaya
Surabaya, Indonesia

Sumit Kumar Sonkar
Department of Chemistry
Malaviya National Institute of Technology
Jaipur, India

Amit Kumar Sonker
Department of Chemical Engineering
and
Department of Materials Science & Engineering
Indian Institute of Technology
Kanpur, India

Ankit Tyagi
Department of Chemical Engineering
Indian Institute of Technology
Kanpur, India

Y. K. Dasan, A. H. Bhat, and Imran Khan

1 Nanocellulose and nanochitin for water remediation by adsorption of heavy metals

1.1 Introduction

The rapidly growing world population and accelerated industrialization have led to a large number of severe environmental problems including water pollution. In terms of the World Health Organization (WHO) drinking water quality guidelines, one-sixth of the population or almost 1.2 billion people living in developing countries are still without access to clean drinking water, while about 2.6 billion people comprising about 400 million children under 5 years old and elderly people do not have access to basic sanitation facilities [1, 2].

In addition, 3.7% of the annual health burden worldwide is caused by unsafe water and lack of sanitation facilities. Among all the water pollutants, heavy metal contaminations are posing a serious threat to human society. Three categories of heavy metals such as toxic metals, precious metals, and radionuclides are of environmental concern. A substantial amount of various toxic metals is released into the water system by many types of industries, such as mining and smelting of minerals, surface finishing industry, energy and fuel production, fertilizer and pesticide industry and application, metallurgy, iron and steel, electroplating, electrolysis, electro-osmosis, leatherworking, photography, electric appliance manufacturing, aerospace and atomic energy installation. For example, mining industries release heavy metal ions such as lead (Pb^{2+}), mercury (Hg^{2+}), silver (Ag^{+}), chromium (Cr^{3+}), arsenic (As^{5+}), cadmium (Cd^{2+}), palladium (Pd^{2+}), zinc (Zn^{2+}), and aluminum (Al^{3+}) to the environment. The recovery of these valuable metal ions after removal is also an issue that needs to be further addressed [3–5]. Silver, copper, and iron are the target metals in this chapter, since all of them belong to the most common pollutants in industrial effluents. Silver ions can be released into groundwater and surface water by many industrial operations such as mining, photographic processing, and electroplating in mirror industry. Industries such as dyeing, paper, petroleum, copper brass plating, and copper–ammonium rayon discharge Cu^{2+} containing wastewater. Short-term exposure to copper ions can result in gastrointestinal distress, and long-term exposure leads to liver or kidney damage [6].

Y. K. Dasan, A. H. Bhat, Department of Fundamental and Applied Sciences, Universiti Teknologi Petronas Malaysia, Perak Darul Ridzuan, Malaysia
Imran Khan, CICECO-Aveiro Institute of Materials, Department of Chemistry, University of Aveiro, Aveiro, Portugal

https://doi.org/10.1515/9783110650600-001

1.2 Current water remediation techniques and their limitation

Numerous conventional methods are readily available for removal of heavy metal ions and improve the effluent quality produced from industrial wastewater such as chemical precipitation, ion exchange, membrane separation, adsorption, electrochemical techniques, and biosorption. In this section, we will discuss about the available treatment methods and their pros and cons.

1.2.1 Chemical precipitation

Chemical precipitation is an effective method for removal of metal ions from wastewater stream by precipitating the metal in an insoluble form. The resulting metal precipitates then settled in a pond and/or a clarifier [7]. The most commonly used chemical precipitation techniques include hydroxide and sulfide precipitation. Chemical precipitation with hydroxide is preferred, as the process involved is relatively simple, is of low cost, and is easy to control the pH. The optimum pH for effective precipitation is in the pH range of 8.0–11.0. In industrial settings, lime is the favored base in hydroxide precipitation [8]. Equation (1.1) represents the conceptual mechanism of heavy metal removal by hydroxide precipitation:

$$M^{n+} + n(OH^-) \longleftrightarrow M(OH)_n \downarrow \tag{1.1}$$

In general, the addition of coagulants such as alum, iron salts, and organic polymers enhances the hydroxide precipitation process [8]. A coagulant destabilizes the suspended particles and makes them flocculate together into larger aggregates that can settle out of solution [9]. On the other hand, metal sulfide precipitates have lower metal ion leach out as compared to hydroxide precipitates. Therefore, sulfide precipitation process has the ability to attain higher treatment efficiency with better thickening and dewatering characteristic sludge. However, this sulfide treatment method produces toxic H_2S fumes due to the acidic nature of heavy metal and sulfide precipitants. Hence, this sulfide precipitation process must be conducted in a neutral or basic medium [8, 10]. Equation (1.2) represents the conceptual mechanism of heavy metal removal by sulfide precipitation [11].

$$M^{2+} + S^{2-} \longrightarrow MS \downarrow \tag{1.2}$$

Carbonate precipitation is well known for calcium hardness removal from water. Recently, the precipitation of metal carbonates was proposed for heavy metal removal. This acts as an alternative technique to reduce the large sludge volume that is produced through hydroxide precipitation. Moreover, it facilitates the settling and filtration process [10]. Equation (1.3) represents the conceptual mechanism of heavy metal removal by carbonate precipitation [11]:

$$M^{2+} + CO_3^{2-} \rightarrow MCO_3 \downarrow \tag{1.3}$$

Despite its simplicity and extensive usage, metal precipitation requires the large amount of chemicals in order to achieve the acceptable level concentration before discharging. Furthermore, precipitation process results in excessive sludge production, which contains toxic metals, thus increasing the cost of sludge disposal. Apart from this, the slow metal precipitation and the long-term environmental impacts of sludge disposal make the process a little tricky [12, 13].

Flotation is a technique of using bubble attachment or carrier to separate solid or dispersed liquids from a liquid phase. The carrier can be activated coal, polymeric resin, mineral particles, or a by-product having good sorption properties [14]. There are five different kinds of flotation processes available: (i) vacuum air flotation, (ii) dispersed air flotation, (iii) electroflotation, (iv) dissolved air flotation, and (v) biological flotation. Among them, dissolved air flotation is the one most frequently used for the removal of heavy metal ions. In dissolved air flotation, metal impurities were separated by foaming in adsorptive bubble separation process [13].

1.2.2 Membrane separation

Membrane separation process is capable of removing suspended solid, organic, and inorganic contaminants from wastewater such as heavy metals. The various types of membrane filtration used in commercial are ultrafiltration (UF), nanofiltration (NF), and reverse osmosis (RO).

1.2.2.1 Ultrafiltration

UF has average pore diameter ranging from 10 to 1,000 Å (1–100 nm) and is recognized as a low-pressure membrane filtration process when compared with NF. UF membrane is capable of capturing microsolutes with molecular weight ranging from 300 to 500,000 Da. However, some of the dissolved metal ions could easily pass through the membrane, which leads to new techniques including micellar-enhanced UF (MEUF) and polymer-enhanced UF (PEUF). During MEUF, small particles are bound or solubilized by the surfactant into large micelle and can be separated by the UF membrane. Various metal cations have been removed using MEUF, including Cd^{2+}, Ni^{2+}, Cs^{+}, Sr^{2+}, Mn^{2+}, Pb^{2+}, Cu^{2+}, and Al^{3+}. MFUF has been considered more economical as it does not require high-pressure membrane and it reduces the cost. However, the most feasible method of various heavy metal removals is PEUF. Metal complexing agents such as polyethyleneimine, polyacrylic acid, diethyl-aminoethyl cellulose, and humic acid are the water-soluble polymers used to complex metal ion and are then separated with PEUF [15–18].

1.2.2.2 Nanofiltration

NF is regarded as a pressure-driven membrane separation process with an intermediate process between UF and RO. This membrane separation process has low operating pressure (7–14 bar) and improved selectivity for mono- and multivalent ions plus relatively low capital and operating cost. NF membrane has pore size of 1 nm with molecular weight cut-off in the range of 100–1,000 Da. It is a favorable technology for rejection of heavy metal ions, including nickel, chromium, copper, and arsenic [16, 18].

1.2.2.3 Reverse osmosis

RO uses pressure to force metal-containing solution through the semipermeable membrane that retains the contaminants on one side and allows the purified fluid to pass through to the other side. It is a separation process of applying pressure in excess of osmotic pressure to force the solvent from high solute concentration pass through the membrane to a region of low solute concentration [19]. RO membrane has pore size of about 0.0005 µm, which allows the removal of particles with molecular weight of greater than 150–250 Da [12]. RO is considered to be the more effective separation process for removal of heavy metal ion from inorganic solutions when compared with UF and NF, with rejection percentage over 97% with metal concentration ranging from 20 to 200 mg/L [13].

1.2.3 Electrochemical treatments

Electrochemical treatment technique is used to treat aqueous metal ion solution by plating out the metal ion on a cathode and is able to recover the metals in the elemental state. The tight environmental law has regained the importance of electrochemical process despite their large capital investment and the expensive electricity supply.

Electrocoagulation (EC) is preferred over traditional coagulation-flocculation method as it is able to trap smallest possible particles in the wastewater treatment. Furthermore, EC generates less volume of sludge, no chemical usage, and less retention time. During the EC process, aluminum or iron electrodes were dissolved electrically to aluminum or iron ions for in situ production of coagulant. Using the principle of electrochemistry, metal ion generation takes place at the anode while hydrogen gas is released from the cathode. The generation of hydrogen gas helps to float and flocculate contaminants in wastewater [8].

Electrodeposition (ED) is an electrochemical technology of electrical current between two electrodes that are immersed in electrolyte (electrolysis) to separate and deposit metal ion on an electrode (cathode). Selectivity of metal ion deposition depends on the current or voltage applied between the electrodes. ED has several gains in terms of cost, safety, sludge produced, versatility, and recovery of pure metal [20, 21].

Electroflotation (EF) is a solid–liquid separation process using tiny hydrogen and oxygen gas bubbles that are generated during water electrolysis to float the contaminant on the water surface. Some study demonstrates 99% of metal ions including iron, copper, nickel, zinc, lead, and cadmium removal using EF [15, 22]. Electrodialysis (ED) uses electric potential to separate ionized contaminants in the solution to pass through an ion exchange membrane. Thin sheet of plastic materials with either anionic or cationic characteristics can be used as membranes. During the separation process in cell compartment, the anions get attracted toward the anode while the cations toward the cathode, crossing the anion exchange and cation exchange membranes [23, 24].

1.2.4 Ion exchange

Ion exchange technique involved an irreversible chemical reaction, where ions present in solution are exchanged with similarly charged ions bound to a stationary solid phase (resin). Reversible ion exchange process is favored, as the ion exchanger can be reused again [25]. In general, there are three types of ion exchangers available: (i) cation exchangers that carry exchangeable cations, (ii) anion exchangers that carry exchangeable anions, and (iii) amphoteric ion exchanger that is able to exchange both cation and anion [26]. Ion exchanger can be either synthetic (polymer resin) or natural solid. However, synthetic resins are preferred due to their effectiveness in the removal of metal ions from the solution. The most common cation exchangers are the strongly acidic resin with sulfonic acid groups and weakly acid resins with carboxylic acid groups. Besides the synthetic resins, naturally occurring silicate minerals such as zeolites have been used widely for heavy metal removal purpose as they are abundant and of low cost [27].

1.2.5 Clay/layered double hydroxides

Natural materials such as clay minerals have a great potential to be used for heavy metal removal from wastewater stream. Clay mineral is readily available with some peculiar properties such as abundant availability, cheapness, high cation exchange capacity and excellent chemical and mechanical stability [12]. Metal ions can be removed by ion exchange or a complexion reaction at the surface of clays. The important cations and anions found on clay surface are Ca^{2+}, Mg^{2+}, H^+, K^+, NH^{4+}, Na^+, SO_4^{2-}, Cl^-, PO^3, and NO^{3-} which can be exchanged with other ions relatively easily without affecting the clay mineral structure. Anions, cations, and nonionic contaminants and polar contaminants absorbed onto the edges and the faces of clay minerals from natural water. Accumulated contaminants on clay mineral surface leads to their immobilization through the processes of ion exchange, coordination, or ion–dipole interactions. Furthermore, pollutants can also be held through H-bonding, van der Waals forces, or hydrophobic bonding arising from either strong or weak interactions [28]. Adsorbents such as layered materials with interlamellar reactivity have also been tried for metal ion removal process as they have good ion exchange capacity and intercalation properties. Layered double hydroxide (LDH) is an anionic clay consisting of positively charged brucite-like layers in the form of $\left[M^{2+}_{1-x}M^{3+}_{x}(OH)_2\right]^{x+}$ with trivalent cations partially substituting for divalent cations. The trivalent cations in the hydroxide layers are compensated by anions as well as water molecules in the interlayer regions, which leads to the generation of excess positive charges on LDH layers, that is, $A^{m-}_{x/m}nH_2O^{x-}$. Therefore, LDH is expressed as $\left[M^{2+}_{1-x}M^{3+}_{x}(OH)_2\right]^{x+}A^{m-}_{x/m}nH_2O$, where M^{2+} represents divalent metal cation, M^{3+} trivalent metal cation, and A^{m-} an anion, and x was denoted as the molar ratio of M^{3+} to the total metal, ranging from 0.15 to 0.33 for pure LDH formation [29–31]. Chemical precipitation and chelation mechanisms were used by LDHs for removal of heavy metal ions. LDHs have been used to decontaminate wastewater with cations such as Mn^{2+}, Fe^{2+}, and Cu^{2+} by precipitation mechanism [12].

1.2.6 Phytoremediation

Phytoremediation is defined as the process of partially or completely remediated contaminant (organic pollutants, radionuclides, and heavy metals) from soil, sludge, sediments, wastewater, and groundwater using plants or associated ones. Remediation of contaminated sites was aided by the variety of plant processes and the physical characteristics of plant. Phytoremediation is an in situ corrective technology driven by solar energy. Furthermore, it is an efficient eco-friendly process and cost effective [32]. However, it has several disadvantages and constraints that restrict its applicability such as it is dependent on the growing condition required by the plant (climate, geology, altitude, and temperature); time taken to remediate sites far exceeds that of other technologies; large-scale operation requires access to agricultural equipment and knowledge; there are possibilities of contaminant collected in sensing tissues released back into the environment; success is dependent on the tolerance of the plant to the pollutant; and contaminants can be collected in woody tissues that are used as fuel [33]. A number of different processes involved in phytoremediation technique include: (i) phytoextraction (toxic metals from the soil were removed by plants into the harvestable part of plants), (ii) phytofiltration (toxic metals from water system were collected by plant roots), (iii) phytostabilization (bioavailable toxic metals from soil were removed by metal-tolerant plants), (iv) phytovolatilization (contaminants from soil were collected by plants and were transformed into volatile form and transpiring them into the atmosphere), and (v) phytodegradation (plant roots were used to collect the organic molecules and break down into simpler molecules to be stored in the plant tissue) [12, 34]. Phytoextraction technology uses plants to remove the toxic metals from soil into the harvestable parts of plants. Various plant species used in phytoremediation practices are water Kentucky bluegrass (*Poa pratensis*), poplar trees (*Populus* spp.), alfalfa (*Medicago sativa*), hyacinths (*Eichornia crassipes*), American pondweed (*Potamogeton nodosus*), forage kochia (*Kochia* spp.), *Scirpus* spp., coontail (*Ceratophyllum demersum* L.), and the emergent common arrowhead (*Sagittaria latifolia*) among others [35].

1.2.7 Photocatalysis

Photocatalysis is relatively a simple reduction or reaction of capturing contaminants (electron), prompt by ultraviolet radiation in the form of photons (<390 nm) and sensitized by photocatalysts such as titanium ions. Furthermore, electrons are promoted from the valence band to the conduction band of semiconducting oxide when TiO_2 is illuminated with light of $\lambda < 390$ nm to give election–hole pairs. Hydroxide radicals generated at the valence band are a powerful oxidizing agent and attacks organic pollutants that are present near the TiO_2 surface. This process is called as a photo-oxidation of contaminants [36]. Equations (1.4), (1.5), and (1.6) represent the photogeneration of electron and hole, and the formation of OH radicals:

$$TiO_2 + h\nu \rightarrow TiO_2 + e^- + h^+ \quad (1.4)$$

$$H_2O + 2h^+ \rightarrow OH^- + H^+ \quad (1.5)$$

$$TiOH + h^+ \rightarrow TiOH^- \quad (1.6)$$

1.2.8 Adsorption

Adsorption process is defined as the accumulation of a gas or a liquid on the surface of a solid or a liquid (adsorbent), forming a molecular or atomic film (adsorbate). The efficiency of sorption process is influenced by factors such as initial metal ion concentration, ionic strength, adsorbent dosage and type, competing metal ion, complexing agent, and background electrolytes [20]. Some of the commercial adsorbents that are worth to cite for are activated carbon, silica, and grapheme [37]. Steps involved in contaminant sorption onto solid sorbent are (i) the transport of contaminants from bulk solution to the sorbent surface; (ii) adsorption of contaminants on the sorbent surface; and (iii) then the contaminant transport within the sorbent material [23].

1.2.8.1 Magnetic nanoparticles and nanosorbents

Development of magnetic nanoparticles and nanosorbents with nanoporous structure has attracted considerable attention in water treatment industries due to the large surface area, large number of surface active sites, and high magnetic properties that contribute to high adsorption efficiency [25, 38, 39]. Nanosorbents that have been utilized for heavy metal removal are nanoalumina, nanomagnet coated by ethylenediaminetetraacetic acid, carbon nanotube, and hydroxyapatite nanoparticle [40]. Magnetic adsorbents can evade of processing large amount of waste volume by applying external magnetic field. Surface modification of bare magnetic particles is desirable as they are sensitive to air oxidation and tend to aggregate in aqueous system [41].

1.2.8.2 Activated carbon adsorption

Adsorption by activated carbon is one of the most established for effective removal of heavy metal ions. Activated carbons exist in the form of granular activated carbon (GAC), powdered activated carbon (PAC), activated carbon cloth (ACC), and activated carbon fibrous (ACF). The PAC form has large pore diameters and a smaller internal surface area, while GAC owns large internal surface area and small pores. On the other hand, both ACC and ACF have a large surface area with comparatively higher percentage of larger pores. Commercial activated carbon (CAC) has been used widely as it has large surface area (800–1,500 m^2/g), pore volume of 0.20–0.60 cm^3/g and high degree of surface reactivity. Activated carbon can be prepared from all carbonaceous materials by pyrolysis process at temperatures lower than 1,000 °C. There are two major steps involved in the preparation of activated carbon: (i) raw materials carbonized in an inert atmosphere with operating temperature below 800 °C; (ii) then activation of carbonized product at temperature between 950 and 1,000 °C [42]. Activated carbon is an effective adsorbent in reducing hexavalent chromium, mercury, and many metals complexed by organic ligands from heavy metal waste stream [20]. However, high cost of CAC limits the usage in industries for heavy metal removal treatments [43, 44].

1.2.8.3 Low-cost biomass biosorption

Recently, the use of agricultural by-products as adsorbent for heavy metal removal has been given a great deal of interest. Biosorption is defined as the use of agriculture and biological wastes in bioremediation of heavy metal ions. Inactivate microbial biomass was utilized in order to bind and concentrate the heavy metal ions solely by physicochemical pathway of uptake. The various agricultural by-product resources that have been used to prepare activated carbon include coconut shell, coal rice husk, palm kernel husk, corncobs, wool, rice hulls, sawdust, modified cellulosic material, fruits peels, and palm fruit bunch [45]. The surface hydroxyl, carboxyl, sulfate, phosphate, and amino groups are responsible for their affinity to bind heavy metal removal [46]. The utilization of agricultural wastes is encouraged as they are low-cost adsorbents with high efficiency of heavy metal ion removal, helps to reduce the generation of chemical and/or biological sludge, can be regenerated, does not need to use additional nutrients, and provides possibilities of metal recovery [45]. Studies reported that agricultural biowaste and natural material-derived adsorbents that are able to remove metal ions include Cd(II), Cr(III), Cr(VI), Cu(II), Ni(II), and Zn(II).

1.3 Polysaccharides for biosorption

Adsorption is a well-known method in water treatment industries for removal of toxicant and to maintain sanitary requirements. Activated carbon is one of the conventional adsorbents for the removal of heavy metal from waste stream. However, the high cost of activated carbon and activation process limits the application of this material in wastewater treatment industries. Therefore, the urge to prepare cheap and effective adsorbent from natural polymer has arisen. The adsorption of heavy metal by nonliving biomass adsorbent is known as biosorption [47–49]. Numerous studies have been conducted in developing cheap and effective metal sorbents using cellulose and chitin which are structural polysaccharides.

Polysaccharides play an important role in nature, as they are abundant, renewable, low cost, and biodegradable. Furthermore, polysaccharides have high potential for adsorption and serve as building blocks in the construction of mechanically strong structures, as energy storage molecules, and as a recognition and signaling intermediate. Polysaccharides can be classified according to their origin such as animal origin (chitosan), microbial origin (dextran), alga origin (alginate), and vegetal origin (pectin). They can also be classified based on their charges such as anionic (hyaluronic acid, heparin), cationic (chitosan), and nonionic (dextran). In general, polysaccharides exhibit hydrophilic groups such as hydroxyl, carboxyl, and amino groups [50]. Therefore, it is necessary to analyze their structure to find out about their functional diversity. Polysaccharides are formed by an arrangement of three or more monosaccharides along the polymer chain, which gives rise to the degree of structural variety. The most common polysaccharides for adsorption processes are cellulose, chitin/chitosan, starch, cyclodextrin, and guar gums, as well as their chemically modified derivatives are now used as an adsorbent for heavy metals and dyes from water [51, 52].

Cellulose is one of the abundant organic compounds that exist on the Earth. Cellulose is defined as a polysaccharide comprising hundreds of glucose units linked in a beta 1-4 configuration. It is a structural component of the cell walls of many plants. Despite the fact that primary source of cellulose is wood, there are other major sources such plant fibers (cotton, hemp, flax, etc.), marine animals (tunicate), or algae, fungi invertebrates, and bacteria. Besides that, agricultural residues such as rice bran, rice husk, wheat husk, wheat bran bark of trees, maize corn cob, sugar cane bagasse, fruit peels, soybean hulls, saw dust, cotton stalks, and sugar beet pulps can be the potential source of cellulose for metal biosorption process. Most of these agricultural biowastes are cheap, abundant, eco-friendly, and potential source of cellulose [53]. There are four major mechanisms involved in the removal of heavy metal by cellulose, including formation of complexes with dissolved cellulose hydrolysis products, adsorption onto the cellulose fibers, intercalation into the cellulose matrix, and formation of chemical bond with the reactive group of cellulose. During the adsorption reaction, hydrated metal ions present in aqueous system swell the cellulose by absorbing into it and the hydrated metal ion shell gets exchanged with OH groups present on the surface of cellulose. Furthermore, carboxylic acid derivatives (carboxylic and hydrocarboxylic acids) formed during the alkaline hydrolysis of cellulose are strong ligands for binding heavy metal ions in alkaline media. In general, metal ions and co-ligands (ammonia, tartaric acid, etc.) in the form of complexes in aqueous media and cellulose as a ligand attract the metal ion when the interaction between metal ion and co-ligand is weakened [54].

Chitin is a biomaterial of animal origin and the second most abundant biopolymer after cellulose. It is one of the natural polysaccharides that carry one amine or amide group per glucose ring (95% nitrogen) [55]. The high content of amino and hydroxyl groups in chitin derivatives in particular chitosan has gained broad attention as it shows a significant adsorption potential for the removal of heavy metal ions [56]. Furthermore, it was reported that chitin can readily take up some metal ions such as iron, zinc, copper, lead, mercury, and uranium (biosorption of heavy metals). Chitins are mainly extracted from exoskeleton of crustaceans farmed/harvested for human consumption, chiefly shrimp, crab, and lobsters. Chitin is one of the promising biomaterial due to the physicochemical, chemical stability, biodegradability, biocompatibility, high reactivity, and nontoxicity [57]. In general, the interaction between metal ion and chitin can occur by different mechanisms, including ion exchange mechanism or chelation of metal ion on the free electronic doublet of amino groups. In chitin/chitosan, nitrogen atom of amino group has one lone pair, while oxygen atom of hydroxyl group has two lone pairs of electron which can bind a metal ion through a lone pair of electrons forming a complex. Metal ions have greater affinity for nitrogen atom than the oxygen atom due to the difference in electronegativity. Hence, the amine groups are responsible for the uptake of metal ion by chelating mechanism [58].

1.4 Nanopolysaccharides

Nanomaterials have attracted the attention of scientists and industrialists as they enable the produced materials with new and improved properties. Nanomaterials are defined as materials having a characteristic length in the range of approximately from 1 to 100 nm. There are numerous possible commercial and technological applications of nanomaterials such as environmental science, electronics, information processing, optoelectronics, catalysis, biomedical science, advanced defense technologies, and energy conversion and storage [59].

Polysaccharide-based nanomaterials including nanocellulose and nanochitin have very promising properties for the broad range of applications. Polysaccharides are the most abundant natural polymer available on the Earth. They possess magnificent structural diversity and functional versatility, which are often superior to synthetic materials [60]. The properties of polysaccharides (cellulose and chitin) can easily be tailored according to the desired application due to the high level of functional groups such as hydroxyl, amino, and carboxylic acid groups [61]. Therefore, numerous studies have been conducted on sorption properties of polysaccharides for heavy metal removal technology. However, the biosorption efficiency of amorphous polysaccharides was limited due to their low mechanical properties [62]. Hence, modification of polysaccharides into nanopolysaccharides is important in order to obtain or improve their characteristics.

Nanopolysaccharides possess excellent properties, including high elastic modulus, improved mechanical properties, high aspect ratio, and high Young's modulus. Nanopolysaccharides that exist in nanometer scale have high surface area with large number of active sites. The high specific surface area has potentials to be used for storage, separation, and purification process. Studies reported that cellulose and chitin nanofiber have Brunauer–Emmett–Teller (BET) surface areas of 480 and 260 m^2/g, respectively [63, 64]. Moreover, nanopolysaccharides can be used in high-pressure environment since they have good mechanical strength and rigidity [65].

Various methods have been used in literatures for the production of nanocellulose and nanochitin, including mechanical, chemical, and enzymatic isolation and chemomechanical processes. Generally, mechanical defibrillation of nanopolysaccharides involves grinding, cryocrushing, and homogenizing. During chemical isolation, the amorphous region of polysaccharides was hydrolyzed by acids such as sulfuric acid, hydrochloric acid, and phosphoric acid under the controlled environment [66]. Enzymatic hydrolysis involved the use of bacteria for selective hydrolysis of specific component that exists in polysaccharide [67].

1.5 Hierarchical structure of cellulose and chitin

Cellulose is a structural plant polymer with repetitive unit $(C_6H_{10}O_5)_n$ that is linked by β-(1 → 4)-glycosidic bonds (Figure 1.1). Length origin of cellulose determines the length of β-(1 → 4) glucan chain. The compact microfibrils of plant are made from ordered manner of closely packed cellulose chains. The inter- and intramolecular hydrogen bond present in cellulose chain creates highly ordered and three-dimensional crystal structure, which make them insoluble in water and in most of the organic solvents. Cellulose exists in four crystalline structures, including cellulose (I, II, III, and IV). Native cellulose that is found in nature is cellulose I, which occurs in two allomorphs,

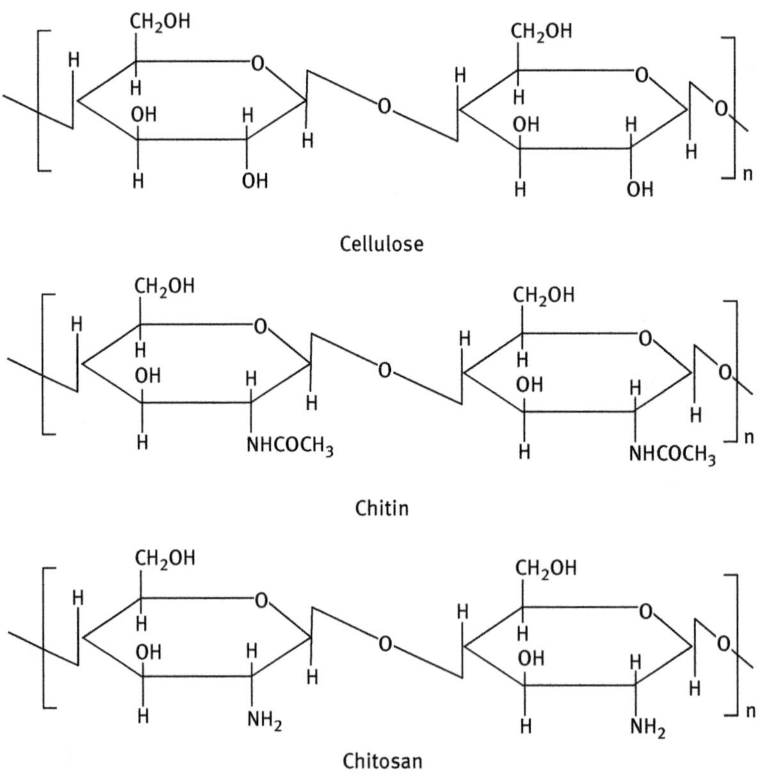

Figure 1.1: Structure of cellulose, chitin, and chitosan.

Iα and Iβ. Regenerated cellulose is produced through recrystallization or mercerization with aqueous sodium hydroxide identified as cellulose II. Cellulose II is thermodynamically the most stable form of crystallites. Cellulose I is distinct from cellulose II due to the layout of their atoms, where cellulose I runs in a parallel direction while cellulose II has antiparallel arrangement. On the other hand, treating cellulose I and II with liquid ammonia leads to the formation of cellulose III1 and III2, respectively. The further modification of cellulose III1 and III2 in glycerol at 206 °C produces cellulose IV1 and IV2, respectively [66, 68, 69].

Chitin is a semicrystalline homopolymer made up of *N*-acetylglucosamine (2-acetomido-2-deoxy-D-glucose, GlcNAc) joined through β-(1,4)-linkage (Figure 1.1). Native chitin occurs in three different polymorphic forms (α, β, and γ). The three different forms of chitin were differentiated in terms of their packing and polarity of adjacent chains in successive sheets: α-chitin chains are all in antiparallel arrangement, β-chitin chains are in parallel arrangement, and γ-chitin has two parallel chains alternate with one antiparallel chain.

There polymeric forms of chitin can be distinguished through infrared spectroscopy, solid-state nuclear magnetic resonance spectroscopy and X-ray diffraction. In nature, α-chitin is the most abundant, which can be obtained from exoskeletons of crustaceans. Further, the most reactive β-chitin extracted from squid pens shows higher affinity for solvents. Fungi and yeast are the major sources of γ-chitin. Chitosan is one of the deacetylated soluble derivatives of chitin [70–75]. The hierarchical structure of wood fiber and exoskeleton of crab is shown in Figure 1.2.

These naturally occurring polysaccharides have a hierarchical organization with various structural level. Polysaccharides (cellulose and chitin) have macro- and nanoscale particles due to their hierarchical structure and semicrystalline nature. Micro- and nanofibrils exist in the natural polymer are small and mechanically stronger with alternating crystalline and amorphous domains. They are mainly referred to as nanocellulose and nanochitin [76].

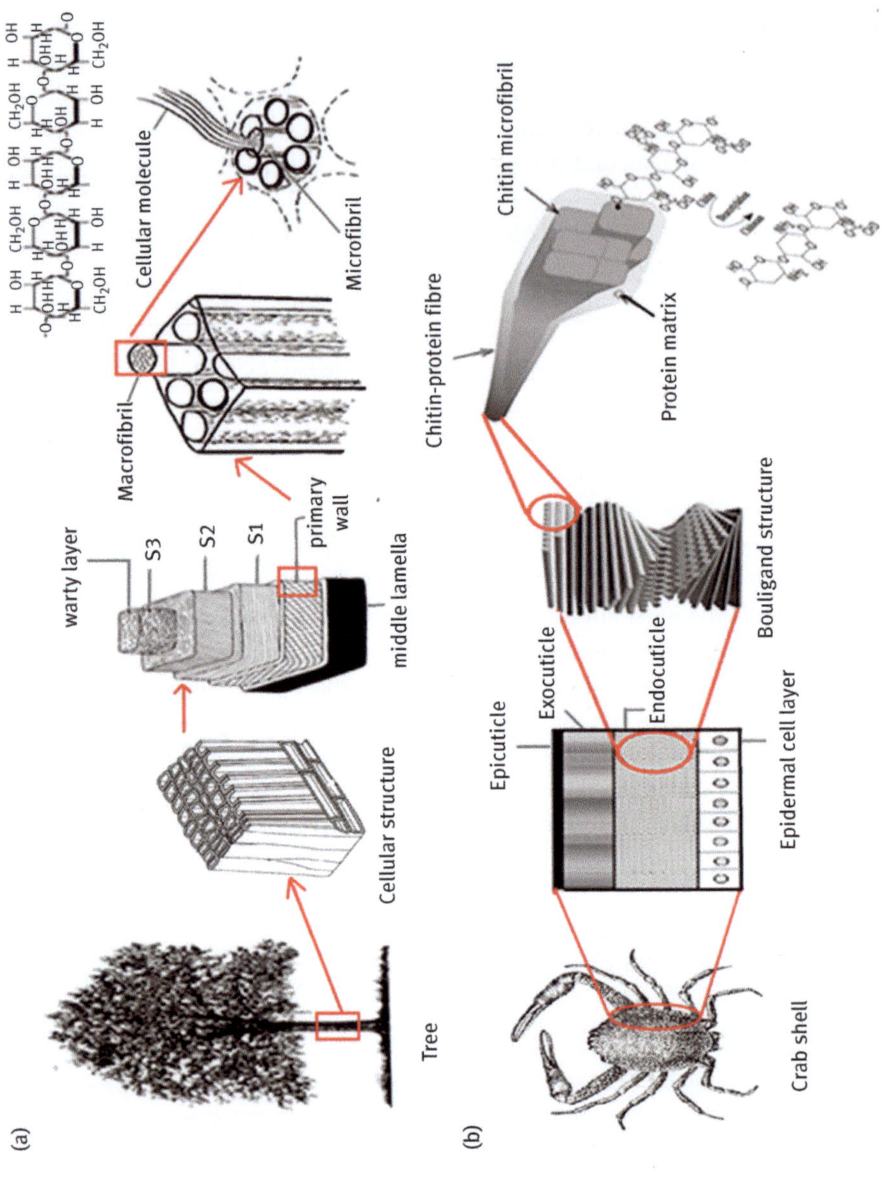

Figure 1.2: Hierarchical structure of (a) wood fiber and (b) exoskeleton of crab.

1.6 Conclusions

The current studies have proved that nanopolysaccharides have potential to adsorb a wide range of heavy metal ions, including Ag(I), Cu(II), Fe(III), Ni(II), Cr(III), and Zn(II), but have varying adsorption velocity and capacity. In general, cellulose nanocrystals have better adsorption behavior than native cellulose nanofibers, which are attributed to the surface functionalities and specific surface area. The metal adsorption behavior of nanopolysaccharides was found to be pH dependent, and the best adsorption performance was observed near neutral pH. Therefore, metal ions immobilized on nanopolysaccharides can be recovered by acidic solution wash. The adsorption behavior of nanopolysaccharides has been dramatically enhanced by chemical modifications with large amount of metal binding groups such as carboxylic groups and phosphoryl groups grafted onto the surface by TEMPO (2,2,6,6-tetramethylpiperidine-1-oxyl radical) oxidation or enzymatic phosphorylation. The nanopolysaccharide adsorbents were found to be highly efficient in model water system as well as in industrial effluents containing multiple heavy metal ions.

The combination of efficient isolation to nanoscale and optimal chemical modifications can be utilized to achieve excellent metal biosorbents at competitive price. The green and abundant aspects of the waste bioresources envisaged in this work, their good performance in terms of heavy metal ions adsorption, and the possibilities to recycle them after use are strong arguments to consider them as sorbents for heavy metal ions removal from contaminated water.

References

[1] Shannon, M.A., Bohn, P.W., Elimelech, M., Georgiadis, J.G., Marinas, B.J., and Mayes, A.M. Science and technology for water purification in the coming decades. Nature, 2008, 452(7185), 301.

[2] Ashbolt, N.J. Microbial contamination of drinking water and disease outcomes in developing regions. Toxicology, 2004, 198(1–3), 229.

[3] Wang, J., and Chen, C. Biosorption of heavy metals by Saccharomyces cerevisiae: a review. Biotechnology Advances, 2006, 24(5), 427.

[4] Moore, J.W., and Ramamoorthy, S. Heavy metals in natural waters: applied monitoring and impact assessment, Springer Science & Business Media, New York, 2012.

[5] Purcell, T.W., and Peters, J.J. Sources of silver in the environment. Environmental Toxicology and Chemistry, 1998, 17(4), 539.

[6] Aksu, Z., Sag, Y., and Kutsal, T. The biosorpnon of copperod by C. vulgaris and Z. ramigera. Environmental Technology, 1992, 13(6), 579.

[7] Wang, L.K., Tay, J.-H., Tay, S.T.L., and Hung, Y.-T. Environmental bioengineering, Springer Science & Business Media, New Jersey, 11, 2010.

[8] Fu, F., and Wang, Q. Removal of heavy metal ions from wastewaters: a review. Journal of Environmental Management, 2011, 92(3), 407.

[9] Pepper, I.L., Gerba, C.P., and Brusseau, M.L. Environmental and pollution science, Academic press, London, UK, 2011.
[10] Wang, L.K., Hung, Y.-T., and Shammas, N.K. Physicochemical treatment processes, Springer, Totowa, New Jersey, 3, 2005.
[11] Guyer, H.H. Industrial processes and waste stream management, John Wiley & Sons, New York; Chichester, 1998.
[12] Sharma, S.K. Heavy metals in water: presence, removal and safety, Royal Society of Chemistry, Cambridge, UK, 2014.
[13] Sarkar, B. Heavy metals in the environment, CRC Press, Boca Raton, Florida, USA, 2002.
[14] Rubio, J., and Tessele, F. Removal of heavy metal ions by adsorptive particulate flotation. Minerals Engineering, 1997, 10(7), 671–679.
[15] Roy, A., and Bhattacharya, J. Nanotechnology in industrial wastewater treatment, IWA Publishing, London, UK, 2015.
[16] Pabby, A.K., Rizvi, S.S., and Requena, A.M.S. Handbook of membrane separations: chemical, pharmaceutical, food, and biotechnological applications, CRC press, Boca Raton, Florida, USA, 2015. 9781466555563.
[17] Mohanty, K., and Purkait, M.K. Membrane technologies and applications, CRC Press, Boca Raton, Florida, USA, 2011. 9780429075483.
[18] Lichtfouse, E., Schwarzbauer, J., and Robert, D. Green materials for energy, products and depollution, Springer, Berlin Heidelberg, 2013, 978-3-540-26531-3.
[19] Bakalár, T., Búgel, M., and Gajdošová, L. Heavy metal removal using reverse osmosis. Acta Montanistica Slovaca, 2009, 14(3), 250–253.
[20] Chen, J.P. Decontamination of heavy metals: processes, mechanisms, and applications, CRC Press, Boca Raton, Florida, USA, 2012, 9780429130939.
[21] Iorga, M.I., Buzatu, D., Mirica, M.C., and Pop, R. Methods for the enhancement of mass transport at the recovery of metal ions from hydroelectrometallurgical processes, INTECH Open Access Publisher, 2012.
[22] Wang, L.K., Hung, Y.-T., and Shammas, N.K. Advanced physicochemical treatment technologies, Springer, 2007.
[23] Barakat, M. New Trends in Removing Heavy Metals from Industrial Wastewater. Arabian Journal of Chemistry, 2011, 4(4), 361.
[24] Kurniawan, T.A., Chan, G.Y., Lo, W.-H., and Babel, S. Physico-chemical treatment techniques for wastewater laden with heavy metals. Chemical Engineering Journal, 2006, 118(1), 83.
[25] Hubicki, Z., and Kołodyńska, D. Selective removal of heavy metal ions from waters and waste waters using ion exchange methods, INTECH Open Access Publisher, 2012.
[26] Luqman, M. Ion exchange technology II: applications, Springer Science & Business Media, 2, 2012.
[27] Ersoz, M., and Barrott, L. Best practice guide on metals removal from drinking water by treatment, IWA Publishing, 2012.
[28] Srinivasan, R. Advances in Materials Science and Engineering, 2011, 2011.
[29] Hu, Q., Xu, Z., Qiao, S., Haghseresht, F., Wilson, M., and Lu, G.Q. A novel color removal adsorbent from heterocoagulation of cationic and anionic clays. Journal of Colloid and Interface Science, 2007, 308(1), 191.
[30] Zhang, M., Gao, B., Yao, Y., and Inyang, M. Phosphate removal ability of biochar/MgAl-LDH ultra-fine composites prepared by liquid-phase deposition. Chemosphere, 2013, 92(8), 1042.
[31] Chen, D., Li, Y., Zhang, J., Zhou, J.-z., Guo, Y., and Liu, H. Magnetic Fe3O4/ZnCr- layered double hydroxide composite with enhanced adsorption and photocatalytic activity. Chemical Engineering Journal, 2012, 185, 120.

[32] Dixit, R., Malaviya, D., Pandiyan, K., Singh, U.B., Sahu, A., Shukla, R., Singh, B.P., Rai, J.P., Sharma, P.K., Lade, H., and Paul D. Bioremediation of Heavy Metals from Soil and Aquatic Environment: An Overview of Principles and Criteria of Fundamental Processes. Sustainability, 2015, 7(2), 2189.

[33] Rungwa, S., Arpa, G., Sakulas, H., Harakuwe, A., and Timi, D. Phytoremediation - An Eco-friendly and Sustainable Method of Heavy Metal Removal from Closed Mine Environments in Papua New Guinea. Procedia Earth and Planetary Science, 2013, 6, 269.

[34] Carels, N., Sujatha, M., and Bahadur, B. Jatropha, challenges for a new energy crop, Springer Science & Business Media, 1, 2012.

[35] Paz-Alberto, A.M. and Sigua, G.C. Phytoremediation: A green technology to remove environmental pollutants. American Journal of Climate Change, 2, 71–86.

[36] Samarghandi, M., Nouri, J., Mesdaghinia, A., Mahvi, A., Nasseri, S., and Vaezi, F. Efficiency removal of phenol, lead and cadmium by means of UV/TiO2/H2O2 processes. International Journal of Environmental Science & Technology, 2007, 4(1), 19.

[37] Karnib, M., Kabbani, A., Holail, H., and Olama, Z. Heavy Metals Removal Using Activated Carbon, Silica and Silica Activated Carbon Composite. Energy Procedia, 2014, 50, 113.

[38] Xu, P., Zeng, G.M., Huang, D.L., Feng, C.L., Hu, S., Zhao, M.H., Lai, C., Wei, Z., Huang, C., and Xie, G.X. Use of iron oxide nanomaterials in wastewater treatment: a review. Science of the Total Environment, 2012, 424, 1.

[39] Giraldo, L., Erto, A., and Moreno-Piraján, J.C. Magnetite nanoparticles for removal of heavy metals from aqueous solutions: Synthesis and characterization. Adsorption, 2013, 19(2–4), 465.

[40] Hao, Y.-M., Man, C., and Hu, Z.-B. Effective removal of Cu (II) ions from aqueous solution by amino-functionalized magnetic nanoparticles. Journal of Hazardous Materials, 2010, 184(1), 392.

[41] Carlos, L. Applications of magnetite nanoparticles for heavy metal removal from wastewater, 2013.

[42] Bansal, R.C., and Goyal, M. Activated carbon adsorption, CRC press, 2005.

[43] Saifuddin, M., and Kumaran, P. Removal of heavy metal from industrial wastewater using chitosan coated oil palm shell charcoal. Electronic Journal of Biotechnology, 2005, 8(1), 43.

[44] Davis, E.B. Focus on environmental research, Nova Publishers, 2006.

[45] Demirbas, A. Heavy metal adsorption onto agro-based waste materials: a review. Journal of Hazardous Materials, 2008, 157(2), 220.

[46] Bulut, Y. Removal of heavy metals from aqueous solution by sawdust adsorption. Journal of Environmental Sciences, 2007, 19(2), 160.

[47] Datskevich, E., Prikhod'ko, R., Stolyarova, I., Lozovskii, A., and Goncharuk, V. Water treatment to remove acid and basic dyes by biosorption on polysaccharide composites. Russian Journal of Applied Chemistry, 2010, 83(10), 1785.

[48] Son, B.C., Park, K., Song, S.H., and Yoo, Y.J. Selective biosorption of mixed heavy metal ions using polysaccharides. Korean Journal of Chemical Engineering, 2004, 21(6), 1168.

[49] Davis, T.A., Volesky, B., and Mucci, A. A review of the biochemistry of heavy metal biosorption by brown algae. Water Research, 2003, 37(18), 4311.

[50] Silva, A.K.A., Letourneur, D., and Chauvierre, C. Polysaccharide Nanosystems for Future Progress in Cardiovascular Pathologies. Theranostics, 2014, 4(6), 579.

[51] Carrara, S. Nano-bio-sensing, Springer Science & Business Media, 2010.

[52] Crini, G., and Badot, P.-M. Sorption processes and pollution: conventional and non-conventional sorbents for pollutant removal from wastewaters, Presses Univ. Franche-Comté, 2010.

[53] Sud, D., Mahajan, G., and Kaur, M. Agricultural waste material as potential adsorbent for sequestering heavy metal ions from aqueous solutions - a review. Bioresource Technology, 2008, 99(14), 6017.

[54] Öztürk, H.B., Vu-Manh, H., and Bechtold, T. Interaction of cellulose withe alkali metal ions and complexed heavy metals. Lenzinger Berichte, 2009, 87, 142.
[55] Volesky, B. Removal and recovery of heavy metals by biosorption. In: Volesky B, editor. Biosorption of Heavy Metals. Boca.Ration, FL: CRC Press. pp. 7–43.
[56] Morghi, M., Abidar, F., Soudani, A., Zerbet, M., Chiban, M., Kabli, H., and Sinan, F. Removal of nitrate ions from aqueous solution using chitin as natural adsorbent. International Journal of Research in Environmental Studies, 2015, 2(2), 8–20.
[57] Sharp, R.G. A Review of the Applications of Chitin and Its Derivatives in Agriculture to Modify Plant-Microbial Interactions and Improve Crop Yields. Agronomy, 2013, 3(4), 757.
[58] Chauhan, S. Modification of chitosan for sorption of metal ions. Journal of Chemical & Pharmaceutical Research, 2015, 7(4).
[59] Yin, Y., and Talapin, D. The chemistry of functional nanomaterials. Chemical Society Reviews, 2013, 42(7), 2484.
[60] Dumitriu, S. Polysaccharides: structural diversity and functional versatility, CRC press, 2004.
[61] Wen, Y., and Oh, J.K. Recent Strategies to Develop Polysaccharide-Based Nanomaterials for Biomedical Applications. Macromolecular Rapid Communications, 2014, 35(21), 1819.
[62] Dumitriu, S. Polysaccharides in medicinal applications, CRC Press, 1996.
[63] Heath, L., Zhu, L., and Thielemans, W. Chitin Nanowhisker Aerogels. ChemSusChem, 2013, 6(3), 537.
[64] Sehaqui, H., de Larraya, U.P., Liu, P., Pfenninger, N., Mathew, A.P., Zimmermann, T., and Tingaut, P. Enhancing adsorption of heavy metal ions onto biobased nanofibers from waste pulp residues for application in wastewater treatment. Cellulose, 2014, 21(4), 2831.
[65] Liua, P., Borrella, P.F., Bozic, M., Kokolb, V., Oksmana, K., and Mathew, A.P. Nanocelluloses and their phosphorylated derivatives for selectiveadsorption of Ag^{+}, Cu^{2+} and Fe^{3+} from industrial effluents. Journal of Hazardous Materials, 2015, 294, 177–185.
[66] Corrêa, A.C., de Morais Teixeira, E., Pessan, L.A., and Mattoso, L.H.C. Cellulose nanofibers from curaua fibers. Cellulose, 2010, 17(6), 1183.
[67] Lee, H., Hamid, S., and Zain, S. Conversion of lignocellulosic biomass to nanocellulose: structure and chemical process. The Scientific World Journal, 2014, 2014.
[68] Lavoine, N., Desloges, I., Dufresne, A., and Bras, J. Microfibrillated cellulose - its barrier properties and applications in cellulosic materials: a review. Carbohydrate Polymers, 2012, 90(2), 735.
[69] Khalil, H.A., Bhat, A., and Yusra, A.I. Green composites from sustainable cellulose nanofibrils: A review. Carbohydrate Polymers, 2012, 87(2), 963.
[70] Zhou, D., Zhang, L., Zhou, J., and Guo, S. Cellulose/chitin beads for adsorption of heavy metals in aqueous solution. Water Research, 2004, 38(11), 2643.
[71] Kumar, M.N.R. A review of chitin and chitosan applications. Reactive and Functional Polymers, 2000, 46(1), 1.
[72] Rinaudo, M. Chitin and chitosan: Properties and applications. Progress in Polymer Science, 2006, 31(7), 603.
[73] Pillai, C., Paul, W., and Sharma, C.P. Chitin and chitosan polymers: Chemistry, solubility and fiber formation. Progress in Polymer Science, 2009, 34(7), 641.
[74] Dutta, P.K., and Dutta, J. Multifaceted development and application of biopolymers for biology, biomedicine and nanotechnology, Springer, 254, 2013.
[75] Ruiz-Herrera, J. Fungal cell wall: structure, synthesis, and assembly, CRC press, 1991.
[76] Grossman, R.F., Nwabunma, D., Dufresne, A., Thomas, S., and Pothan, L.A. Biopolymer nanocomposites: processing, properties, and applications, John Wiley & Sons, 8, 2013.

Kumud Malika Tripathi, Ankit Tyagi, Amit Kumar Sonker, and Sumit Kumar Sonkar

2 Waste-derived nanocarbons: a cleaner approach toward water remediation

2.1 Introduction

Water remediation to obtain clean water from deteriorated water is currently a huge challenge for the global scientific concern [1–4]. Rapid and continuous growth of global population and industrialization is directly or indirectly contributing a huge volume of waste material to the environment [5–8]. Therefore, to control this, an ecological development toward waste minimization via exploration of newer methods utilizing waste materials is highly desirable [9–12]. For example, direct conversion of waste/low-cost carbon materials to nanocarbons with different morphologies such as carbon nanotubes (CNTs) [13–15], carbon nano-onions [16, 17], graphene [18, 19], carbon nanofibers (CNFs) [20], carbon nanospheres (CNSs) [10, 21], carbon nanocubes [22], photoluminescent carbon dots (CDs) [23, 24], graphene quantum dots (GQDs) [25, 26], and carbon aerogels (materials with highest known porosity, surface area, three-dimensional structures with superhydrophobicity) [27, 28] has also been reported. The frequent uses of nanocarbons are emerging as sustainable strategy for the utilization of maximum wastes [17, 22, 23, 29] as used in water purification [30–35], in killing pathogens [21, 36], as sensors for biomolecules and heavy metals [37–39], in bioimaging [40–44], as fertilizer [17, 45], and in the field of photoconversion [46, 47]. Synthesis/isolation of nanocarbons from globally generated black carbon (BC) particulate matter known as BC offers a significant approach for the worldwide reduction of BC, thus directly decreasing the extent of environmental pollution [16, 32, 48, 49]. In common practice, carbon nanoparticles (CNPs) are discharged into the atmosphere continuously by various natural and anthropogenic processes such as automobiles, industries and by burning wood and oil lamp/candles, which increases air pollution in both indoor and outdoor atmosphere [15, 23, 49–51]. Another potential use of waste-derived nanocarbons, related to purification of water as its cleaner availability directly reflects the global health.

Kumud Malika Tripathi, Smart Plastics Group, University of South Brittany (UBS), LIMATB-UBS, Lorient, France
Ankit Tyagi, Department of Chemical Engineering, Indian Institute of Technology, Kanpur, India
Amit Kumar Sonker, Department of Materials Science & Engineering, Indian Institute of Technology, Kanpur, India
Sumit Kumar Sonkar, Department of Chemistry, Malaviya National Institute of Technology, Jaipur, India

https://doi.org/10.1515/9783110650600-002

As described in literature, exposure of CNPs to humans has been known since ages [50]. Carbon has been used since the historic age in the form of charcoal for the purification of drinking water [19]. Charcoal is found to be very efficient for the removal of impurities and contaminants from water via chemical adsorption [19]. Presently, CNPs are emerged as high impact materials for water remediation as a potential alternative for activated carbon (AC) [19, 36, 52]. Current advancements in the field of carbon nanotechnology have attracted enormous stimulation and concern in almost every specialized/interdisciplinary field of science and technology [53]. Advances in research and development of carbon-based nanomaterials have paved the way as smart material for water remediation as low cost and green alternative to conventionally used materials [52, 54]. Because of the presence of unique physiochemical characteristic properties such as high surface area to volume ratio by which they have a high tendency for adsorption [52, 54], along with tunable surface functionalities and biocompatibility [33, 55–58]. However, such CNPs produced by chemical methods such as chemical vapor deposition, arc discharge, and plasma-assisted synthesis have long processing times and complex operation steps and cytotoxicity issues [59–61], limiting their environmental application. Another hindrance to CNPs industrial adaption is their high fabrication cost. In this regard, waste/biowaste-derived CNPs assumed as a cost-effective alternative could lead to a game-changing transformation in future for water remediation.

Herein, we focus on the latest advancement in CNP-based water treatment technologies, applications in practical world, and future prospective. This chapter provides an overview on the current and latest progress for the exploration of synthetic advances of CNPs from such resources with wastes/biowastes. The recent research in the field of sensing, adsorption, photodegradation, and antimicrobial activity of CNPs is discussed. The significant developments and progress in the field of water remediation with waste-derived CNPs are reported.

2.2 Waste materials as a "carbon precursor" for the synthesis of CNPs

As stated earlier, utilization of waste material for the development of newer cost-effective CNPs is a promising approach toward waste management. Suriani et al. reported the synthesis of quasi-aligned CNTs by the waste engine oil [62]. In another report, they demonstrated the synthesis of CNFs from waste chicken fat [63]. Many more reports were available for the waste-derived synthesis of CNTs, such as by using plastic waste [64] (polypropylene [65, 66] polyethylene, polyethylene terephthalate [64]), waste of printed circuit board [67], waste of red mud [68], bioethanol waste [69], paraffin wax soot [70], soot from burning grass [71], waste of cooking oil [72], and waste from glycerol [73]. A considerable volume of literature is available for the high yield synthesis of Carbon dots (CDs) from waste and waste biomasses. For instance, CDs are efficiently synthesized by numerous methods using biochar [74], polystyrene foam waste [16], shrimp egg [75], petroleum coke [76], waste frying oil [77], vegetable extract [78], date molasses [79], waste paper [80], rice husk [81], citrus pectin [82], kerosene soot [83], lime juice [84], apple juice [85], bee pollens [86], goose feathers [87], sugarcane juice [88], and many more as waste/low-cost carbon source [43, 89–91]. Fabrication of carbon aerogels are diverse in nature and can be fabricated from a variety of precursor materials, including organic compounds, inorganic complexes, petroleum, and bioresources [27, 28, 92]. Sun et al. reported the combined methodology by using microwave and hydrothermal techniques for the synthesis of fluorinated GQDs from graphene waste [93]. Adolfsson et al. derived GQDs from biowaste (cellulose) via the formation of CNSs as an intermediate product by the microwave-assisted hydrothermal degradation of cellulose [25]. Based on the as-discussed numerous simple method, the conversion of waste to CNPs can be extended to industrial scale with the hope to utilize its dual prospects: waste minimization along with water remediation.

2.3 Advancements in water remediation using waste-derived CNPs

Plant soot and waste material-derived ACs are highly used materials for water remediation due to highly activated surface area, pore size, and low cost as well [94, 95]. Adsorptions associated with ACs are currently being used as mature technology for the removal of many soluble chemical contaminates [96]. But, in addition to purification and separation purposes, reuse of absorbent materials after remediation is also an inherent limitation and frequently resulted in degradation of the adsorption properties [11]. Alternatively, magnetic mesoporous CNPs have been reported for wastewater treatment due to their uniform architecture of pores, and respectable chemical and thermal stability [31]. In contrast to AC, magnetic CNPs can be easily recycled and regenerated from treated medium by the application of magnetic field. CNPs synthesized by using the various waste/biowaste materials offer a variety of alternates with fast and economic greener techniques [19, 57, 97]. CNPs with tuned surface functionalities hold a great potential for the removal of engineered nanoparticles (NPs) and other organic/inorganic contaminants, dissolved toxic metals, and so on [98, 99]. Synthesis and designing of multifunctional CNPs for an operational, real-time, fast, efficient, and accurate water remediation to remove diverse type of contaminates is still a challenge.

2.3.1 Sensing/removal of soluble toxic metal ions and other inorganic pollutants

With the increase in urbanization and industrialization, transition metal ions turn into a serious concern because of their potential biological and environmental hazards [100, 101]. It is renowned that transition metal ions play a crucial role in various physicochemical and environmental processes such as catalysis, biosynthesis, and transportation but not all [102], few transition metal ions such as Hg^{2+}, Cr^{6+}, As^{3+}, and Cd^{2+} are well known for their potential carcinogenic and mutagenic impacts [100, 103]. Among all, Hg^{2+} is one of the most dangerous and ubiquitous metals due to its ability to cross the biological membrane and bioaccumulated inside the cellular system causing serious damage to endocrine, central nervous system, and mitotic impairment and decreasing the process of photosynthesis and transpiration in plants [101]. Cd^{2+} can cause metabolism disorder, cancer risk assessment, mutagenesis, and renal dysfunction [100, 104]. Cu^{2+} is potentially toxic to living organisms due to oxidative damage/interference of copper homeostasis [105], causing Alzheimer's, Menkes, Prion, and Wilson's diseases [106]. An increase in the concentration of free Al^{3+} is very lethal for growing plants [107, 108].

Conventional analytical approaches for the detection of trace amounts of toxic metal ions include the ultraviolet–visible spectrometry, atomic absorption/emission spectroscopy, inductively coupled plasma mass spectrometry, auger electron spectroscopy, and polarography [38]. However, sophisticated instruments and complex process of sample preparation with the use of expensive chemicals constrain their easy practical applications [38]. Thus, the development of economic and efficient analytical technique with high sensitivity and accuracy is desired. In this regard, fluorescent-based sensors have been fascinated with extensive interest to detect metal ions and other inorganic pollutants both in vivo and in vitro [39, 102, 106].

The engagement of CNPs especially CDs and GQDs has been emerged as a promising candidate for the detection of heavy metal ion based on their unique fluorescence "switch off"/"switch on" mechanism [109]. Shi et al. synthesized dual-emission CDs from naked oat and used them for the selective sensing of Al^{3+} [110]. Gaddam et al. described the synthesis of blue-green fluorescent CDs from camphor soot [37], and used these for the detection of Cd^{2+}, Hg^{2+}, Zn^{2+}, Fe^{2+}, and Cu^{2+} in aqueous solution via fluorescence quenching [37]. Lu et al. synthesized CNPs from pomelo peel via hydrothermal treatment. The synthesized CNPs were well separated and are in the range of 2–4 nm in diameter as demonstrated in Figure 2.1(a) and (b) [111]. These CNPs were further employed for the sensitive and selective detection of Hg^{2+} by the fluorescence quenching method with the detection limit of 0.23 nm. Strong emission of CNPs at 444 nm decreases with the addition of Hg^{2+} and effectively quenches in intensity as demonstrated in Figure 2.1(c). The fluorescence was restored on the addition of strong chelating agent cysteine (Figure 2.1(c) curve c) since Hg^{2+} was removed from the CNP surface and strongly bind to cysteine by the formation of Hg–S bond. The selectivity of

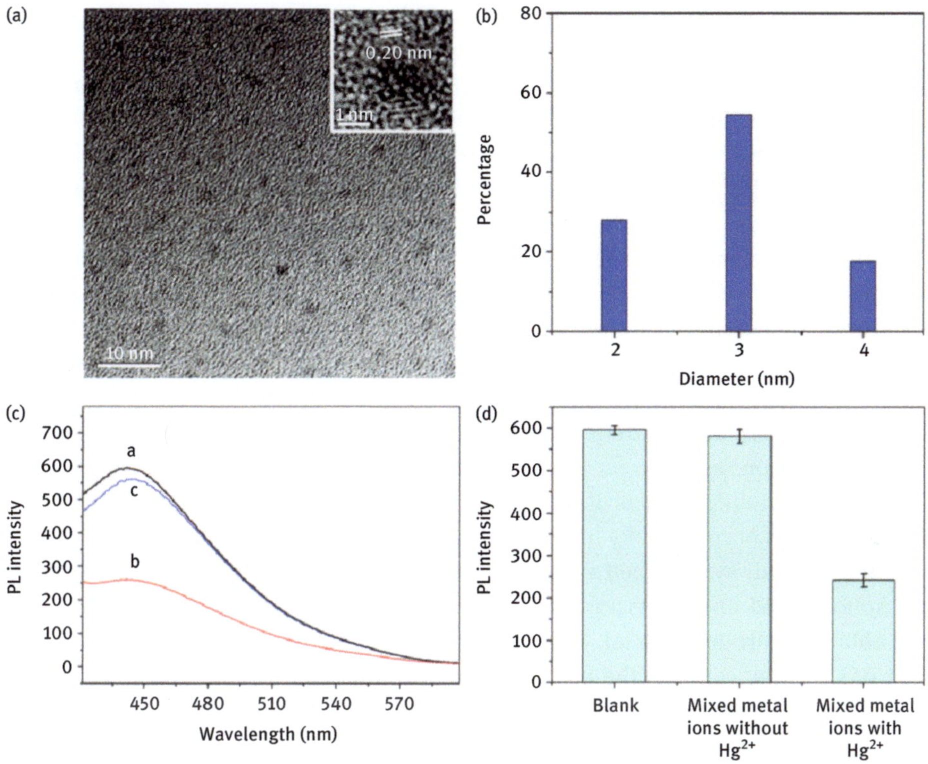

Figure 2.1: (a) TEM images of the products thus formed; the inset shows the HRTEM image. (b) Corresponding particle size distribution histogram, (c) PL spectra of solutions of CPs (curve a), CPs-Hg^{2+} mixture (curve b), and CPs-Hg^{2+-} Cys mixture (curve c). The difference in PL intensity of CPs dispersion under various conditions in PBS (pH 7.0; excitation at 360 nm; mixed metal ions including Ag^{+}, Ca^{2+}, Cd^{2+}, Co^{2+}, Cu^{2+}, Fe^{3+}, Hg^{2+}, Mg^{2+}, Mn^{2+},Ni^{2+}, Pb^{2+}, and Zn^{2+}; $[M^{n+}] = 5 \times 10^4$ nM). Reproduced with permission from W. Lu, X. Qin, S. Liu, G. Chang, Y. Zhang, Y. Luo, A. M., Asiri, A. O. Al-Youbi, and X., Sun, Analytical Chemistry, 2012, 84, 12, 5351. ©2012 American Chemical Society [111].

CNPs in the presence of other metal ions is shown in Figure 2.1(d). CNPs show a good selectivity toward Hg^{2+} in the presence of other interference ions (5×10^4 nM). They proposed that strong binding affinity of Hg^{2+} with surfacial carboxylic acid group than other metal ion attributed good selectivity. They also checked the efficiency of CNPs for real water sample obtained from a lake and found to be effective for real-time detection.

Tan et al. demonstrated an efficient way for the synthesis of fluorescent CDs from a food industry waste (sago waste) by thermal pyrolysis [112]. The pyrolysis condition greatly affects the fluorescent behavior of synthesized CDs. At 250 °C carbonization temperature, CDs obtained were larger in size (~37–66 nm in diameter) with spherical morphology. On the other hand, at higher temperature of pyrolysis at 400 °C, particle size was comparatively small (~6–17 nm in diameter). They also observed that an increase in carbonization temperature shifts the emission trends to blue shift with decrease in intensity at 315 nm excitation wavelength. CDs were further valuated for the fluorescence-based selective sensing of metal ions (Cu^{2+} and Pb^{2+}). The emission intensity efficiently quenched with an increase in Cu^{2+} and Pb^{2+}.

Suryawanshi et al. reported an easy synthetic methodology for the formation of Graphene quantum dots (GQDs) from biowastes and further surface functionalized these with ammonia solution to amine-terminated (Am-GQDs) for selective sensing of Ag^+ ions [113]. High-resolution transmission electron microscopy (HRTEM) characterization of GQDs and Am-GQDs are shown in Figure 2.2(a) and (b), respectively. They used fluorescence on–off–on technique for the detection of Ag^+ in aqueous solution. A schematic representation for proposed mechanism is demonstrated in Figure 2.2(e). Fluorescence quenching was observed by various metal ions but on addition of cysteine, restoration occurred only in Ag^+ (Figure 2.2(c) and (d)).

Yeung et al. synthesized activated CNPs from coffee waste in the presence of phosphoric acid and potassium hydroxide in order to increase the surface area of the obtained carbon material. The specific surface area was found to be 677.2 ± 2.7 and 708 ± 0.1 $m^2\ g^{-1}$ after the impregnation with phosphoric acid and KOH, respectively, via the methylene blue (MB) adsorption method. They found that the CNPs thus obtained are not toxic to *Escherichia coli* and human HaCaT skin cells. They further used these CNPs for the effective removal of Cu^{2+} and Pb^{2+} [114]. Tang et al. synthesized attapulgita/carbon (APT/C) nanocomposite by calcining the Spent bleaching earth (SBE) at different temperatures (250, 300, 350, 400, 450, and 500 °C). SBE is a waste obtained from palm oil industries, in the process of removing the unnecessary components and to improve the quality of palm oil. The APT/C nanocomposite thus obtained was used for the adsorptive removal of Pb^{2+}, Cd^{2+}, and Cu^{2+} from aqueous solution (Figure 2.3). The maximum adsorption capacity was found to be 32.32, 46.72, and 105.61 mg g^{-1} for Cu^{2+}, Cd^{2+}, and Pb^{2+} ions, respectively [115]. Hu et al. reported the synthesis of CNPs from the waste plastic bags (WPBs) collected from dustbins. Hydrothermal treatment of WBPs was done (in autoclave) in the presence of different amounts of H_2O_2 to convert WPBs into CNPs in various sizes depending on the amount of H_2O_2.

They found that on increasing the amount of H_2O_2, smaller size CNPs were formed. They proposed that at higher concentration of H_2O_2, polyethylene is cleaved into smaller oxidized chains. CNPs are highly fluorescent and are used for the selective sensing of Fe^{3+} ions as well as for the bioimaging of Hela cells. They found that CNPs can detect Fe^{3+} ions up to lower concentration of 2.8 μM effectively and selectively quantify Fe^{3+} ions from 10 to 400 μM concentration [116].

For CNSs, Li et al. reported the synthesis of nanocomposite based on CNS adsorbents via hydrothermal carbonization, which was further explored for the removal of

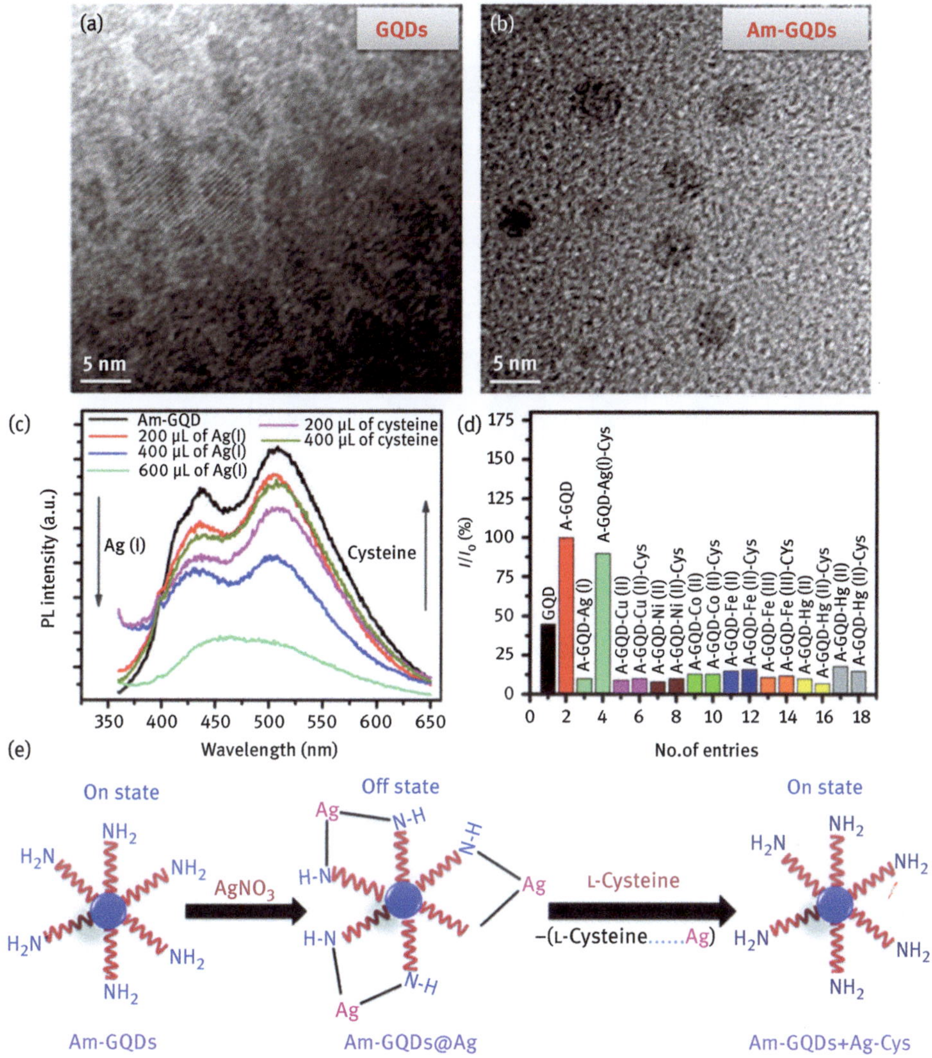

Figure 2.2: HRTEM images of (a) GQDs and (b) Am-GQDs, (c) PL data of quenching upon addition of $AgNO_3$ solution (1 mM) and photoluminescence regeneration upon addition of cysteine (1 mM), (d) selectivity of the on–off–on sensing in metal ions. I_0 is considered as the PL intensity of Am-GQDs and I is the PL intensity of other entries. (e) The suggested on–off–on mechanism of PL sensing of Ag^+ ions by Am-GQDs. Reproduced with permission from A. Suryawanshi, M. Biswal, D. Mhamane, R. Gokhale, S. Patil, D. Guin and S. Ogale, Nanoscale, 2014, 6, 20, 11664. © 2014 The Royal Society of Chemistry [113].

Cr^{6+} from wastewater with an excellent performance of removing 156.25 mg/g of Cr^{6+} [117]. Konicki used hollow CNSs for the efficient removal of anionic dye, non-ionic dye, and cationic dye such as Direct Red 23 (DR23), Direct Green 97 (DG97),

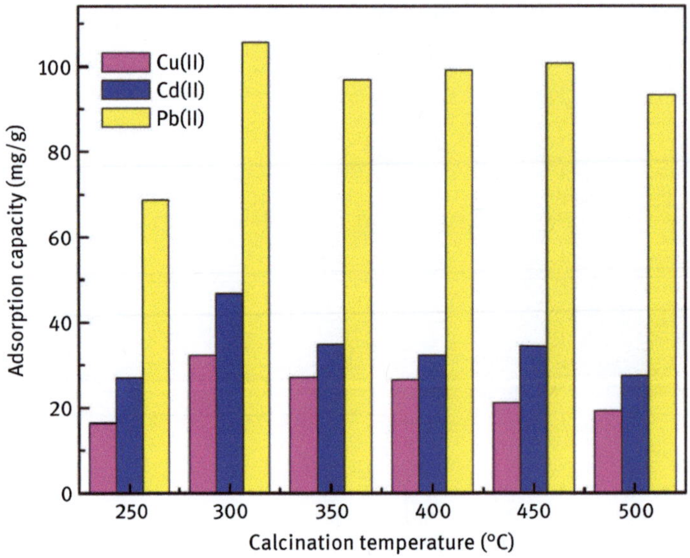

Figure 2.3: Effect of calcination temperature on adsorption capacity of Cu^{2+}, Cd^{2+}, and Pb^{2+}. Reproduced with permission from J. Tang, B. Mu, M. Zheng, and A. Wang, ACS Sustainable Chemistry & Engineering, 2015, 3, 6, 1125. © 2015 American Chemical Society [115].

and Basic Yellow 28 (BY28), respectively, from the aqueous solution via the direct adsorption of these dyes onto the hollow CNSs [118]. Li et al. reported the potential of porous CNSs for the removal of toxic arsenate and selenate from three different water sources: deionized water, well water, and canal water [119]. Song et al. reported the synthesis of colloidal CNSs for selective detection of Ag^+ from aqueous solutions. Importantly, the used CNSs were easily reactivated by using dilute acid solutions [120].

2.3.2 Sensing/removal of organic pollutants and NPs

As we understand, water pollution rises as a consequence of oil spills, human activities, leakage of toxic chemicals, and discharge of industrial effluents [121, 122]. In order to clean up the oil contamination from water, biodegradation, physical diffusion via the addition of dispersants, in situ combustion, and oil skimmer are commonly used techniques [123–125]. However, all these techniques suffer some limitations such as need of precious equipment, lack of selectivity, and generation of other contaminates/toxic by-products [124]. Recently, porous carbon-based superhydrophobic nanomaterials (aerogels) have been used as oil absorber since they offer a simple, economic, fast, and efficient way for oil–water separation due to their specific surface properties [28, 121, 126, 127]. Li et al. synthesized low density and superhydrophobic carbon aerogels via hydrothermal carbonization of green carbon precursor as "winter melon" [28]. The synthesized carbon aerogels exhibited up to 50 times absorption capacity of their weight for organic solvents and oil from water. Zhang and coworkers reported the use of raw cotton as a precursor for the green synthesis of carbon aerogels through inert pyrolysis for the removal of oil and organic solvents from water via absorption [126].

Not the oil only, wastewater from various industries such as printing, textile, plastic, dyestuff, foodstuffs, and cosmetics contains colored organic substances [128], which are highly soluble, aromatic, nondegradable, and very difficult to remove. Contamination of dye even at very low concentration highly influenced the water quality parameter due to its potential toxicity and carcinogenic nature [129]. As well, in addition to this presence of organic pollutants inhibited the sunlight to enter into waterbodies [36]. To remove these, absorption and photocatalytic degradation are highly used and most appealing techniques [130, 131]. CNPs with diverse morphologies have been widely applied as an economical adsorbent of organic pollutant from water [128, 132].

Unur et al. demonstrated the adsorption of MB dye by mesoporous nanocarbon synthesized via hydrothermal technique by using a biowaste as hazelnut shells [97]. The synthesized mesoporous (1,700 $m^2 g^{-1}$) nanocarbon showed higher adsorption rate and efficiency for MB removal in comparison with ACs. The large surface area and acidic surface functionality with the presence of heteroatom (O and N) increased the adsorption affinity for basic dyes from aqueous solution.

Graphene has been extensively used for water remediation purposes [19, 33, 133]. It was reported that when graphene or graphene oxide anchored on sand surface in river, their efficiency for the removal of organic and inorganic pollutants significantly increases [33]. Gupta et al. synthesized graphenic material anchored on river sand surface by using common sugar termed as graphene sand composite (GSC) [19]. GSC was found to be effective for the removal of organic contaminations such as dyes (rhodamine 6G) and pesticides (chloropyrifos) from water via adsorption. Most importantly, GSC can be easily regenerated by column separation with acetone after water purification (Figure 2.4). Sun et al. used the waste-derived GQDs from the pyrolysis of citric acid in aqueous solutions and used these for the selective detection of phenols in water [134].

Wu et al. fabricated the hollow and highly porous carbon spheres from sewage sludge and polystyrene spherical foam (2.0, 3.5, and 4.8 mm in diameter) as template via pyrolysis in inert atmosphere at 500 °C [32], and used for the removal of organic dyes with high efficiency. Magnacca et al. reported the synthesis of magnetic NPs, wherein core–shell of Fe_3O_4 was coated with soluble biobased products (SBO, isolated from urban solid wastes) in the size range of 10 nm [135]. Magnetic NPs were synthesized by modified coprecipitation method with SBO. HRTEM images of synthesized NPs with 0.5 g SBO are demonstrated in Figure 2.5

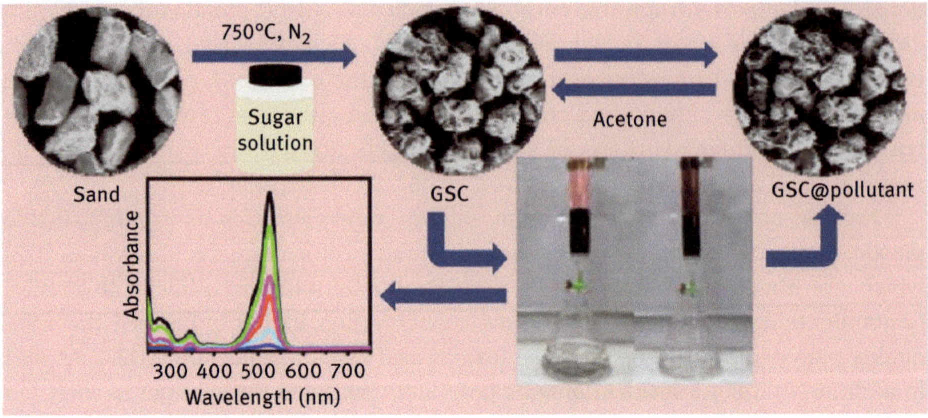

Figure 2.4: Synthesis of graphene from sugar and river sand and application for water remediation. Reproduced with permission from S. S. Gupta, T. S. Sreeprasad, S. M. Maliyekkal, S. K. Das and T. Pradeep, ACS Applied Materials & Interfaces, 2012, 4, 8, 4156. © 2012 American Chemical Society [19].

(a) and (b). HRTEM shows the spherical morphology of NPs coated with organic layer. Hysteresis loop of magnetic NPs at room temperature is shown in Figure 2.5(c), which shows the superparamagnetic nature of NPs, with zero coercivity and remanence. These Fe_3O_4 core and SBO shell NPs for the adsorptive removal of cationic azo dye crystal violet (CV) are also considered as mitotic poison. Most of the removal take place in 30 min, but for the better establishment of sorption equilibrium contact time was chosen as 3 h. The amount of CV removed with the different amount of SBO coated over Fe_3O_4 was given in Figure 2.5(d). Figure 2.5(e) represents aqueous dispersion of NPs and separation in magnetic field. The effect of adsorbent dose was also shown in Figure 2.5(f). They found that almost 95% of the CV was removed at the adsorbent dose of 1,000 mg L^{-1} [135].

Recently, Tripathi et al. demonstrated the adsorption of MB on wsCDs surface via fluorescence quenching technique [23]. Water soluble carbon dots (wsCDs) were synthesized using global pollutant waste diesel engine exhaust waste soot as carbon precursor. Surfacial acidic groups introduced via acid oxidation are proposed to increase the adsorption affinity for basic dyes. MB adsorbed on wsCDs surface via ion pair interaction and efficiently quenched the fluorescence emission as demonstrated in Figure 2.6.

Not always the soluble toxic metal ions, NPs because its small sized are also capable to cross biological barriers and easily translocated within cellular environments and seems to be a serious concern for both present and future [30, 99, 136]. Cayuela et al. proposed a simple approach for the sensing of carboxylated-CNTs via CDs in spiked river water based on the fluorescence quenching technique [30]. Kumar et al. reported the removal of Ag and Au NPs via adsorption method by the hydrothermally

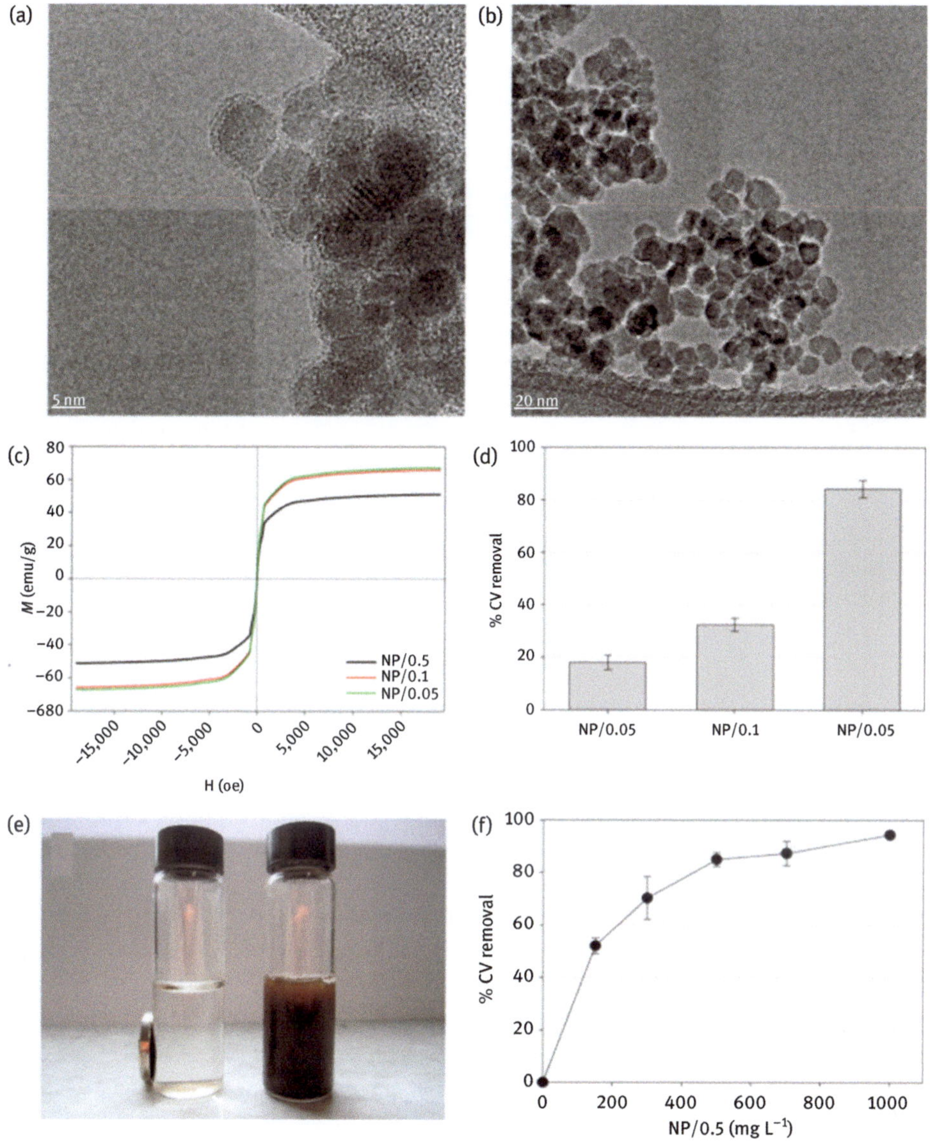

Figure 2.5: (a), (b) HRTEM images of NP/0.5; (c) magnetization curves of NP/0.5, NP/0.1, and NP/0.05 at 300 K; (d) CV removal obtained with NPs prepared with different amounts of SBO. $[CV]_0 = 10$ mg L^{-1}, $[NP]_0 = 500$ mg L^{-1}, pH = 7, $T = 25 \pm 2$ °C; (d) NP/0.5 dispersion in aqueous solution and magnetic separation; (f) effect of NP/0.5 dosage on the adsorption of CV. $[CV]_0 = 10$ mg L^{-1}, pH = 7, $T = 25 \pm 2$ °C. The error bars represent the standard deviation of the data obtained by triplicate. Reproduced with permission from G. Magnacca, A. Allera, E. Montoneri, L. Celi, D. E. Benito, L. G. Gagliardi, M. C. Gonzalez, D. O. Mártire and L. Carlos, ACS Sustainable Chemistry & Engineering, 2014, 2, 6, 1518. © 2014 American Chemical Society [135].

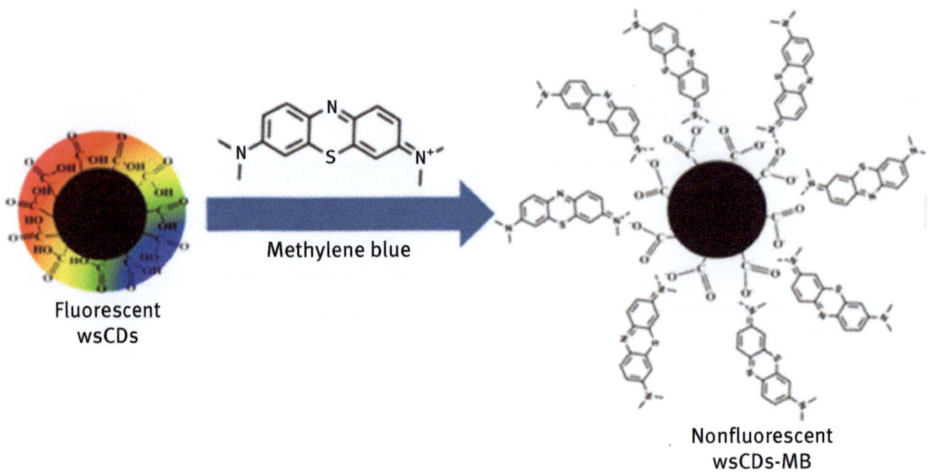

Figure 2.6: A schematic representation for the mechanism of adsorption of MB on wsCDs surface via ion–dipole interaction. Reproduced with permission from K. M. Tripathi, A. K. Sonker, S. K. Sonkar and S. Sarkar, S. RSC Advances, 2014, 4, 57, 30100. © 2014 The Royal Society of Chemistry.

synthesized CNPs (using glucose as a precursor material), which were further passivated with polyethylenimine [57].

2.3.3 Antimicrobial treatment by CNPs

Literature review showed that the commonly used water disinfection techniques are proved to be very efficient toward pathogens. But the formation of toxic byproducts during water remediation has revealed a dilemma [137]. CNPs due to their smaller size and exceptional surface properties hold a great potential as an alternate antimicrobial agent during water remediation processes [138, 139]. Unlike conventional chemical disinfectants, CNPs do not have any toxic byproducts because of their high aqueous stability and chemical inertness [140, 141]. In addition to chemical inertness, CNPs synthesized from wastes exhibited an additional advantage of being economic and green in nature.

The antibacterial activity of soot-generated CNPs are known since ancient times. For example, in Indian culture, pyrolyzing the mustard oil and collecting its soot in a traditional way were used as antimicrobial agents to the eyes and are assumed to prevent ocular infections. Sarkar and coworkers explored the presence of multiwalled CNTs in soot synthesized by burning mustard oil [14]. Saxena et al. investigated the application of water-soluble CNPs synthesized from wood wool for the dose-dependent inhibition of mosquito (*Anopheles*, *Aedes*, and *Culex*) growth from larval to adult stages [56]. Mohanty et al. [142] showed the antimicrobial activity of CNPs collected from mustard oil lamp. Existence of rod-like morphology in lamp soot was observed when dispersed in CCl_4. *Klebsiella pneumoniae* ATCC 29655 and *Staphylococcus aureus* ATCC 13709 were used for the antimicrobial activity. CNPs used here were showing the significant antimicrobial activity toward *K. pneumoniae*, but negligible activity toward *S. aureus*. Varghese et al. separated CNPs from carbon soot collated by the burning of biomass and investigated their antibacterial activity against *Proteus refrigere*, *Streptococcus haemolyticus*, *S. aureus*, and *Pseudomonas aeruginosa* [143].

Purkayastha et al. investigated the second life utilization of high protein content oil industry waste rapeseed oil cake or press-cake for the synthesis of CNPs [21, 144]. Carbon microspheres were synthesized from aqueous solution of CNPs by dropwise addition of acetone in 6:1 ratio. CNPs self-assembled via liquid droplet emulsion method (on addition of acetone) to micrometer size carbon spheres ranging from 160 to 266 nm in diameter. They also synthesized protein-loaded carbon microspheres from purified rapeseed protein. Protein solution was extracted from detoxified rapeseed oil cake. The protein solution (in 0.01 N NaOH) was added in aqueous CNPs solution, and precipitate was separated out by the addition of acetone. CNPs and carbon microspheres show the radical scavenging behavior confirmed by the DPPH-radical scavenging activity. They proposed that radical scavenging behavior and antioxidative potential of CNPs and carbon microspheres due to high surface to volume ratio and surface functionalities. Antimicrobial activity of CNPs and carbon microspheres was analyzed for 10 different types of food-borne pathogens: *Candida albicans* (MTCC183), *S. aureus* (MTCC 3160), *Listeria monocytogenes* AMDK2 (KF894986), *Pseudomonas diminuta* (MTCC 3361), *Bacillus cereus* AMDK1 (KC683896), *Yersinia enterocolitica* (MTCC 859), *K. pneumoniae* (MTCC 618), *Salmonella enterica Typhimurium* (MTCC1252), *E. coli* (MTCC 40), and *Mycobacterium smegmatis* (ATCC 14468). Figure 2.7(a) demonstrates the mean zone of inhibition, which clearly shows the activity of CNPs with the tested microbes. CNPs showed good microbial activity toward *E. coli*, *S. aureus*, *C. albicans*, *P. diminuta*, *S. enterica Typhimurium*, and *Y. enterocolitica*; moderate activity toward *B. cereus*, *K. pneumoniae*, and *L. monocytogenes*; and less microbial activity toward *M. smegmatis*. Carbon microspheres demonstrated almost negligible microbial activity toward all species due to water insolubility and larger size. Figure 2.7(b) and (c) illustrates that TEM was used to characterize the interaction between CNPs and bacteria and also to investigate antimicrobial mechanism by seeing the morphological changes after interaction. In Figure 2.7(d), HRTEM image shows the formation of "pits" on bacterial surface. The degree of plasmid DNA condensation was analyzed by agarose gel electrophoresis, and fluorescent bands of lane 1 are shown in Figure 2.7(e).

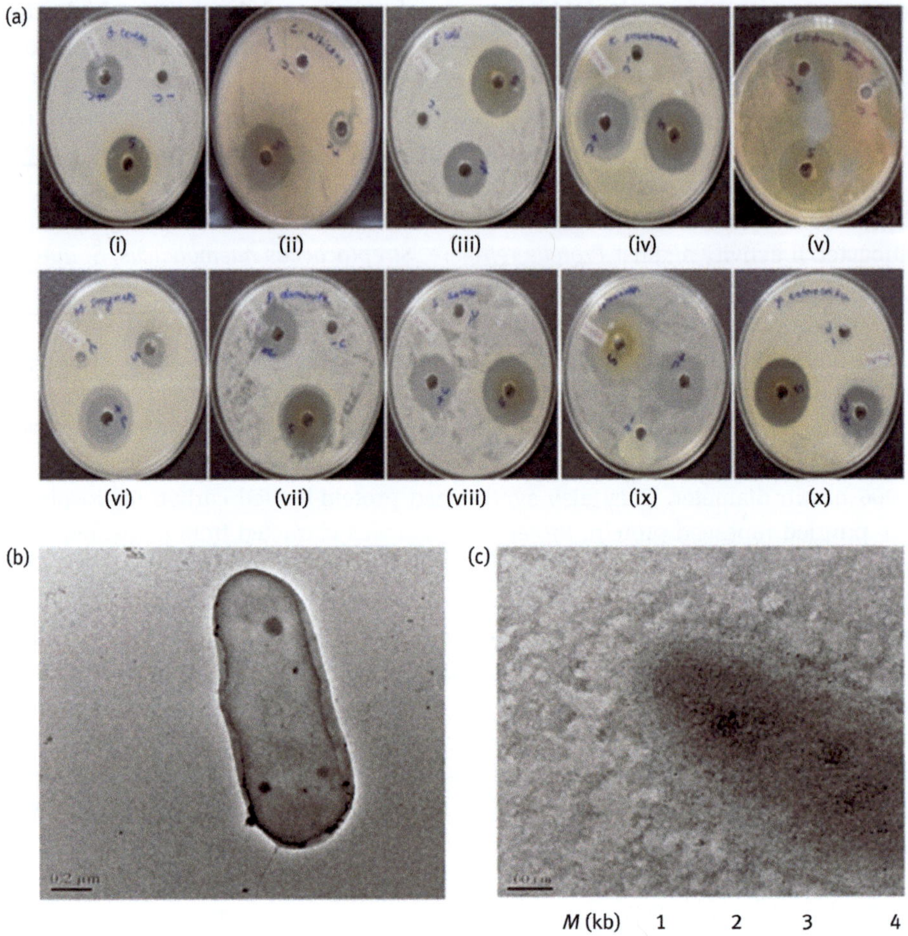

Figure 2.7: (a) (i) *B. cereus* AMDK1 (KC683896), (ii) *C. albicans* (MTCC 183), (iii) *E. coli* (MTCC 40), (iv) *K. pneumoniae* (MTCC 618), (v) *L. monocytogenes* AMDK2 (KF894986), (vi) *M. smegmatis* (ATCC 14468), (vii) *P. diminuta* (MTCC 3361), (viii) *S. aureus* (MTCC 3160), (ix) *S. enterica Typhimurium* (MTCC 1252), and (x) *Y. enterocolitica* (MTCC 859). (−c) and (+c) represent water as negative control and gentamicin sulfate (for bacteria) or nystatin (for fungi) as positive control. (s) indicates sample (CNP), (b) untreated and (c) CNP-treated *E. coli* viewed under TEM, (d) HRTEM image of CNP-treated *E. coli* shows the pit formation in cell membrane, and (e) electrophoretic mobility of CNP–plasmid DNA complex. Reproduced with permission from M. D. Purkayastha, A. K. Manhar, M. Mandal and C. L. Mahanta, Journal of Applied Chemistry, 2014, 2014, 0, 171427. © 2014 Manashi Das Purkayastha et al [21].

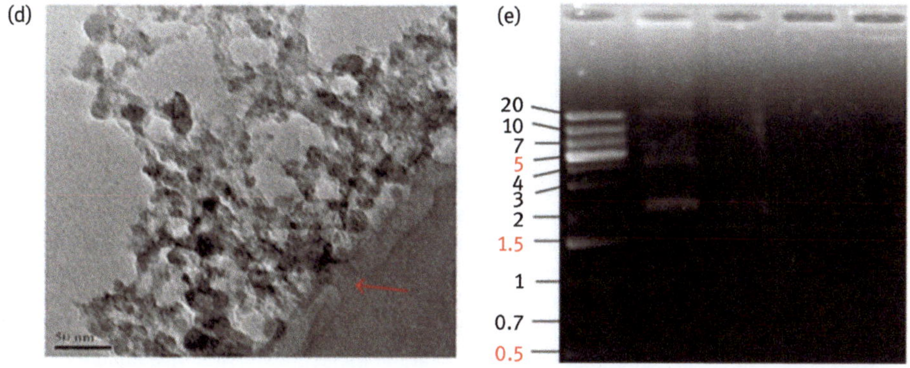

Figure 2.7 (continued)

2.4 Conclusion and future outlook

From the last few years, direct conversion of waste materials to some useful products has increased greatly, especially for the synthesis of carbon-based nanomaterials with different shaped morphologies showing large variety of applications. Among all the existing potential applications, these waste-derived nanocarbons could be easily used for water treatment as this technique is being used since ages. These waste-derived CNPs are expected to contribute a fast, green, and a vast economical solution for drinking water as they are of low cost and are effective adsorbent nanomaterials.

Acknowledgments: K.M.T. acknowledge the support from the NRF grant funded bythe Korea government (NRF- 2019R1G1A1008879). A. T. and A. K. S thank IIT Kanpur for senior teaching assistantship, S. K. S thanks DST New Delhi for funding (SB/EMEQ-383/2014).

References

[1] Vadahanambi, S., Lee, S.-H., Kim, W.-J., and Oh, I.-K. Arsenic removal from contaminated water using three-dimensional graphene-carbon nanotube-iron oxide nanostructures. Environmental Science & Technology, 2013, 47(18), 10510.

[2] Khin, M. M., Nair, A. S., Babu, V. J., Murugan, R., and Ramakrishna, S. A review on nanomaterials for environmental remediation. Energy & Environmental Science, 2012, 5(8), 8075.

[3] Li, Q., Mahendra, S., Lyon, D. Y., Brunet, L., Liga, M. V., Li, D., and Alvarez, P. J. J. Antimicrobial nanomaterials for water disinfection and microbial control: potential applications and implications. Water Research, 2008, 42(18), 4591.

[4] Mishra, A., and Clark, J. Green materials for sustainable water remediation and treatment, The Royal Society of Chemistry, 2013, 1.

[5] Nowack, B., Ranville, J. F., Diamond, S., Gallego-Urrea, J. A., Metcalfe, C., Rose, J., Horne, N., Koelmans, A. A., and Klaine, S. J. Potential scenarios for nanomaterial release and subsequent alteration in the environment. Environmental Toxicology and Chemistry, 2012, 31(1), 50.

[6] Klaine, S. J., Alvarez, P. J. J., Batley, G. E., Fernandes, T. F., Handy, R. D., Lyon, D. Y., Mahendra, S., McLaughlin, M. J., and Lead, J. R. Nanomaterials in the environment: Behavior, fate, bioavailability, and effects. Environmental Toxicology and Chemistry, 2008, 27(9), 1825.
[7] Musee, N. Nanowastes and the environment: Potential new waste management paradigm. Environment International, 2011, 37(1), 112.
[8] Banerjee, S., Tripathi, S. N., Das, U., Ranjan, R., Jadhav, N., Singh, V. P., Jariwala, C., Sonkar, S., and Sarkar, S. Enhanced persistence of fog under illumination for carbon nanotube fog condensation nuclei. Journal Applied Physics, 2012, 112(2), 24901.
[9] Jiang, J., Zhu, J., Ai, W., Fan, Z., Shen, X., Zou, C., Liu, J., Zhang, H., and Yu, T. Evolution of disposable bamboo chopsticks into uniform carbon fibers: a smart strategy to fabricate sustainable anodes for Li-ion batteries. Energy & Environmental Science, 2014, 7(8), 2670.
[10] Hassanzadeh, S., Aminlashgari, N., and Hakkarainen, M. Microwave-Assisted Recycling of Waste Paper to Green Platform Chemicals and Carbon Nanospheres. ACS Sustainable Chemistry & Engineering, 2015, 3(1), 177.
[11] Gupta, V. K., Gupta, B., Rastogi, A., Agarwal, S., and Nayak, A. Pesticides removal from waste water by activated carbon prepared from waste rubber tire. Water Research, 2011, 45(13), 4047.
[12] Laufenberg, G., Kunz, B., and Nystroem, M. Transformation of vegetable waste into value added products: (A) the upgrading concept; (B) practical implementations. Bioresource Technology, 2003, 87(2), 167.
[13] Dubey, P., Sonkar, S. K., Majumder, S., Tripathi, K. M., and Sarkar, S. Isolation of water soluble carbon nanotubes withnetwork structure possessing multipodal junctions andits magnetic property. RSC Advances, 2013, 3(20), 7306.
[14] Dubey, P., Muthukumaran, D., Dash, S., Mukhopadhyay, R., and Sarkar, S. Synthesis and characterization of water-soluble carbon nanotubes from mustard soot. Pramana, 2005, 65(4), 681.
[15] Sonkar, S. K., Tripathi, K. M., and Sarkar, S. Ferromagnetic Behaviour of Anthropogenic Multi-Walled Carbon Nanotubes Trapped in Spider Web Indoor. Journal of Nanoscience and Nanotechnology, 2014, 14(3), 2532.
[16] Dubey, P., Tripathi, K. M., and Sonkar, S. K. Gram scale synthesis of green fluorescent watersoluble onion-like carbon nanoparticles fromcamphor and polystyrene foam. RSC Advances, 2014, 4(12), 5838.
[17] Sonkar, S. K., Roy, M., Babar, D. G., and Sarkar, S. Water soluble carbon nano-onions from wood wool as growth promoters for gram plants. Nanoscale, 2012, 4(24), 7670.
[18] Primo, A., Atienzar, P., Sanchez, E., Delgado, J. M., and Garcia, H. From biomass wastes to large-area, high-quality, N-doped graphene: catalyst-free carbonization of chitosan coatings on arbitrary substrates. Chemical Communications, 2012, 48(74), 9254.
[19] Gupta, S. S., Sreeprasad, T. S., Maliyekkal, S. M., Das, S. K., and Pradeep, T. Graphene from sugar and its application in water purification. ACS Applied Materials & Interfaces, 2012, 4(8), 4156.
[20] Suriani, A. B., Dalila, A. R., Mohamed, A., Mamat, M. H., Salina, M., Rosmi, M. S., Rosly, J., Md Nor, R., and Rusop, M. Vertically aligned carbon nanotubes synthesized from waste chicken fat. Materials Letters, 2013, 101, 61.
[21] Das Purkayastha, M., Manhar, A. K., Mandal, M., and Mahanta, C. L. Industrial waste-derived nanoparticles and microspheres can be potent antimicrobial and functional ingredients. Journal of Applied Chemistry, 2014, 2014, 171427.
[22] Sonkar, S. K., Saxena, M., Saha, M., and Sarkar, S. Carbon nanocubes and nanobricks from pyrolysis of rice. Journal of Nanoscience and Nanotechnology, 2010, 10(6), 4064.
[23] Tripathi, K. M., Sonker, A. K., Sonkar, S. K., and Sarkar, S. Pollutant soot of diesel engine exhaust transformed to carbon dots for multicoloured imaging of E. coli and sensing cholesterol. RSC Advances, 2014, 4(57), 30100.

[24] Babar, D. G., Sonkar, S. K., Tripathi, K. M., and Sarkar, S. P_2O_5 assisted green synthesis of multicolor fluorescent water soluble carbon dots. Journal of Nanoscience and Nanotechnology, 2014, 14(3), 2334.
[25] Adolfsson, K. H., Hassanzadeh, S., and Hakkarainen, M. Valorization of cellulose and waste paper to graphene oxide quantum dots. RSC Advances, 2015, 5(34), 26550.
[26] Roy, P., Periasamy, A. P., Chuang, C., Liou, Y.-R., Chen, Y.-F., Joly, J., Liang, C.-T., and Chang, H.-T. Plant leaf-derived graphene quantum dots and applications for white LEDs. New Journal of Chemistry, 2014, 38(10), 4946.
[27] Pierre, A. C., and Pajonk, G. M. Chemistry of aerogels and their applications. Chemical Reviews, 2002, 102(11), 4243.
[28] Li, Y.-Q., Samad, Y. A., Polychronopoulou, K., Alhassan, S. M., and Liao, K. Carbon aerogel from winter melon for highly efficient and recyclable oils and organic solvents absorption. ACS Sustainable Chemistry & Engineering, 2014, 2(6), 1492.
[29] Angelis, C. D'., Barbosa, do E. S., Corrêa, J. R., Medeiros, G. A., Barreto, G., Magalhães, K. G., de Oliveira, A. L., Spencer, J., Rodrigues, M. O., and Neto, B. A. D. Carbon Dots (C-dots) from Cow Manure with Impressive Subcellular Selectivity Tuned by Simple Chemical Modification. Chemistry – A European Journal, 2015, 21(13), 5055.
[30] Cayuela, A., Soriano, M. L., and Valcárcel, M. Photoluminescent carbon dot sensor for carboxylated multiwalled carbon nanotube detection in river water. Sensors and Actuators B: Chemical, 2015, 207, Part A, 596.
[31] Zhang, Y., Xu, S., Luo, Y., Pan, S., Ding, H., and Li, G. Synthesis of mesoporous carbon capsules encapsulated with magnetite nanoparticles and their application in wastewater treatment. Journal of Materials Chemistry, 2011, 21(11), 3664.
[32] Wu, Z., Kong, L., Hu, H., Tian, S., and Xiong, Y. Adsorption performance of hollow spherical sludge carbon prepared from sewage sludge and polystyrene foam wastes. ACS Sustainable Chemistry & Engineering, 2015, 3(3), 552.
[33] Gao, W., Majumder, M., Alemany, L. B., Narayanan, T. N., Ibarra, M. A., Pradhan, B. K., and Ajayan, P. M. Engineered graphite oxide materials for application in water purification. ACS Applied Materials & Interfaces, 2011, 3(6), 1821.
[34] Sarkar, S., Sonkar, S. K. and Ghoshal, S., ed. I. patent, India, 2009.
[35] Sarkar, S., Allam, A., and Allam, I., US Patent Applications, 2012, p. US 20120012522 A1.
[36] Girgis, B. S., Attia, A. A., and Fathy, N. A. Potential of nano-carbon xerogels in the remediation of dye-contaminated water discharges. Desalination, 2011, 265(1–3), 169.
[37] Gaddam, R. R., Vasudevan, D., Narayan, R., and Raju, K. V. S. N. Controllable synthesis of biosourced blue-green fluorescent carbon dots from camphor for the detection of heavy metal ions in water. RSC Advances, 2014, 4(100), 57137.
[38] Gao, X., Lu, Y., Zhang, R., He, S., Ju, J., Liu, M., Li, L., and Chen, W. One-pot synthesis of carbon nanodots for fluorescence turn-on detection of Ag^+ based on the Ag^+-induced enhancement of fluorescence. Journal of Materials Chemistry C, 2015, 3(10), 2302.
[39] Gogoi, N., Barooah, M., Majumdar, G., and Chowdhury, D. Carbon dots rooted agarose hydrogel hybrid platform for optical detection and separation of heavy metal ions. ACS Applied Materials & Interfaces, 2015, 7(5), 3058.
[40] LeCroy, G. E., Sonkar, S. K., Yang, F., Veca, L. M., Wang, P., Tackett, Kenneth N., Yu, J.-J., Vasile, E., Qian, H., Liu, Y., Luo, P. G., and Sun, Y.-P. Toward structurally defined carbon dots as ultracompact fluorescent probes. ACS Nano, 2014, 8(5), 4522.
[41] Ghosh, M., Sonkar, S. K., Saxena, M., and Sarkar, S. Life Cycle Imaging: Carbon Nano-onions for Imaging the Life Cycle of Drosophila Melanogaster (Small 22/2011). Small, 2011, 7(22), 3170.

[42] Begum, A., Tripathi, K. M., and Sarkar, S. Water-Induced Formation, Characterization, andPhotoluminescence of Carbon Nanotube-Based Composites ofGadolinium(III) and Platinum(II) Dithiolenes. Chemistry – A European Journal, 2014, 20(50), 16657.
[43] Luo, P. G., Yang, F., Yang, S.-T., Sonkar, S. K., Yang, L., Broglie, J. J., Liu, Y., and Sun, Y.-P. Carbon-based quantum dots for fluorescence imaging of cells and tissues. RSC Advances, 2014, 4(21), 10791.
[44] Luo, P. G., Yang, S. S.-T., Sonkar, S. K., Wang, J., Wang, H., LeCroy, G. E., Caoa, L., and Sun, Y.-P. Carbon "quantum" dots for optical bioimaging. Journal of Materials Chemistry B, 2013, 1 (16), 2116.
[45] Tripathi, S., Sonkar, S. K., and Sarkar, S. Growth stimulation of gram (Cicer arietinum) plant by water soluble carbon nanotubes. Nanoscale, 2010, 3(3), 1176.
[46] Wang, J., Sahu, S., Sonkar, S. K., Tackett II, K. N., Sun, K. W., Liu, Y., Maimaiti, H., Anilkumar, P., and Sun, Y.-P. Versatility with carbon dots–from overcooked BBQ to brightly fluorescent agents and photocatalysts. RSC Advances, 2013, 3(36), 15604.
[47] Sahu, S., Cao, L., Meziani, M. J., Bunker, C. E., Shiral Fernando, K. A., Wang, P., and Sun, Y.-P. Carbon dioxide photoconversion driven by visible-light excitation of small carbon nanoparticles in various configurations. Chemical Physics Letters, 2015, 634, 122.
[48] Zribi, K., Feller, J. F., Elleuch, K., Bourmaud, A., and Elleuch, B. Conductive polymer composites obtained from recycled poly (carbonate) and rubber blends for heating and sensing applications. Polymers for Advanced Technologies, 2006, 17(9–10), 727.
[49] Sonkar, S. K., Tripathi, S., and Sarkar, S. Activation of aerial oxygen to superoxide radical by carbon nanotubes in indoor spider web trapped aerosol. Current Science, 2009, 97(8), 1227.
[50] Murr, L. E., and Guerrero, P. A. Carbon nanotubes in wood soot. Atmospheric Science Letters, 2006, 7(4), 93.
[51] Jung, H. S., Miller, A., Park, K., and Kittelson, D. B. Carbon nanotubes among diesel exhaust particles: real samples or contaminants? Journal of the Air & Waste Management Association, 2013, 63(10), 1199.
[52] Kurniawan, T. A., Sillanpää, M. E. T., and Sillanpää, M. Nanoadsorbents for Remediation of Aquatic Environment: Local and Practical Solutions for Global Water Pollution Problems. Critical Reviews in Environmental Science and Technology, 2011, 42(12), 1233.
[53] Fernandes, D., Krysmann, M. J., and Kelarakis, A. Carbon dot based nanopowders and their application for fingerprint recovery. Chemical Communications, 2015, 51(23), 4902.
[54] Shannon, M. A., Bohn, P. W., Elimelech, M., Georgiadis, J. G., Mariñas, B. J., and Mayes, A. M. Science and technology for water purification in the coming decades. Nature, 2008, 452(7185), 301.
[55] Ong, Y. T., Ahmad, A. L., Zein, S. H. S., and Tan, S. H. A review on carbon nanotubes in an environmental protection and green engineering perspective. Brazilian Journal of Chemical Engineering, 2010, 27, 227.
[56] Saxena, M., Sonkar, S. K., and Sarkar, S. Water soluble nanocarbons arrest the growth of mosquitoes. RSC Advances, 2013, 3(44), 22504.
[57] Kumar, J., Mallampati, R., Adin, A., and Valiyaveettil, S. Functionalized carbon spheres for extraction of nanoparticles and catalyst support in water. ACS Sustainable Chemistry & Engineering, 2014, 2(12), 2675.
[58] Roy, M., Sonkar, S. K., Tripathi, S., Saxena, M., and Sarkar, S. Non-toxicity of water soluble multi-walled carbon nanotube on Escherichia-coli. colonies Journal of Nanoscience and Nanotechnology, 2012, 12(3), 1754.
[59] Jia, G., Wang, H., Yan, L., Wang, X., Pei, R., Yan, T., Zhao, Y., and Guo, X. Cytotoxicity of carbon nanomaterials: single-wall nanotube, multi-wall nanotube, and fullerene. Environmental Science & Technology, 2005, 39(5), 1378.

[60] Wang, L., Luanpitpong, S., Castranova, V., Tse, W., Lu, Y., Pongrakhananon, V., and Rojanasakul, Y. Carbon nanotubes induce malignant transformation and tumorigenesis of human lung epithelial cells. Nano Letters, 2011, 11(7), 2796.
[61] Wang, K., Gao, Z., Gao, G., Wo, Y., Wang, Y., Shen, G., and Cui, D. Systematic safety evaluation on photoluminescent carbon dots. Nanoscale Research Letters, 2013, 8(1), 1.
[62] Suriani, A. B., Alfarisa, S., Mohamed, A., Isa, I. M., Kamari, A., Hashim, N., Mamat, M. H., Mohamed, A. R., and Rusop, M. Quasi-aligned carbon nanotubes synthesised from waste engine oil. Materials Letters, 2015, 139, 220.
[63] Suriani, A. B., Dalila, A. R., Mohamed, A., Mamat, M. H., Salina, M., Rosmi, M. S., Rosly, J., Nor, R., and Rusop, M. Vertically aligned carbon nanotubes synthesized from waste chicken fat. Materials Letters, 2013, 101, 61.
[64] Bazargan, A., and Kay, G. M. A review – Synthesis of carbon nanotubes from plastic wastes. Chemical Engineering Journal, 2012, 195–196, 377.
[65] Bajad, G. S., Tiwari, S. K., and Vijayakumar, R. P. Synthesis and characterization of CNTs using polypropylene waste as precursor. Materials Science and Engineering, 2015, 194, 68.
[66] Nahil, M. A., Wu, C., and William, P.T. Influence of metal addition to Ni-based catalysts for the co-production of carbon nanotubes and hydrogen from the thermal processing of waste polypropylene. Fuel Processing Technology, 2015, 130, 46.
[67] Quan, C., Li, A., and Gao, N. Synthesis of carbon nanotubes and porous carbons from printed circuit board waste pyrolysis oil. Journal of Hazardous Materials, 2010, 179, 911.
[68] Oliveira, A. A. S., Tristão, J. C., Ardisson, J. D., Dias, A., and Lagoa, R. M. Production of nanostructured magnetic composites based on Fe^0 nuclei coated with carbon nanofibers and nanotubes from red mud waste and ethanol. Applied Catalysis B: Environmental, 2011, 105, 163.
[69] Alvesa, J. O., Zhuoa, C., Levendis, Y. A., and Tenórioc, J. A. S. Catalytic conversion of wastes from the bioethanol production into carbon nanomaterials. Applied Catalysis B: Environmental, 2011, 106, 433–444, 433.
[70] Li, Y. Y., and Hsieh, C.C. Synthesis of carbon nanotubes by combustion of a paraffin wax candle. Micro & Nano Letters, 2007, 2, 63.
[71] Kang, Z., Wang, E., Mao, B., Su, Z., Chen, L., and Xu, L. Obtaining carbon nanotubes from grass. Nanotechnology, 2005, 16, 1192.
[72] Suriani, A.B., Nor, R. M., and Rusop, M. Vertically aligned carbon nanotubes synthesized from waste cooking palm oil. Journal of the Ceramic Society of Japan, 2010, 118, 963.
[73] Wu, C. ang, Z., William, P. T., and Huang, J. Renewable hydrogen and carbon nanotubes from biodiesel waste glycerol. Scientific Reports, 2013, 3.
[74] Saxena, M., Maity, S., and Sarkar, S. Carbon nanoparticles in 'biochar'boost wheat (Triticum aestivum) plant growth. RSC Advances, 2014, 4, 39948.
[75] Lin, P.-Y., Hsieh, C.-W., Kung, M.-L., Chu, L.-Y., Huang, H.-J., Chen, H.-T., Wu, D.-C., Kuo, C.-H., Hsiehh, S-L., and Hsieh, S. Eco-friendly synthesis of shrimp egg-derived carbon dots for fluorescent bioimaging. Journal of Biotechnology, 2014, 189, 114.
[76] Wu, M., Wang, Y., Wu, W., Hu, C., Wang, X., Zheng, J., Li, Z., Jiang, B., and Qiu, J. Preparation of functionalized water-soluble photoluminescent carbon quantum dots from petroleum coke. Carbon, 2014, 78, 480.
[77] Hu, Y., Yang, J., Tian, J., Jia, L., and Yu, J.-S. Waste frying oil as a precursor for one-step synthesis of sulfur-doped carbon dots with pH-sensitive photoluminescence. Carbon, 2014, 77, 775.
[78] Wang, J., Ng, Y. H., Lim, Y.-F., and Ho, G. W. Vegetable-extracted carbon dots and their nanocomposites for enhanced photocatalytic H_2 production. RSC Advances, 2014, 4(83), 44117.

[79] Das, B., Dadhich, P., Pal, P., Srivas, P. K., Bankoti, K., and Dhara, S. Carbon nanodots from date molasses: new nanolights for the in vitro scavenging of reactive oxygen species. Journal of Materials Chemistry B, 2014, 2(39), 6839.

[80] Wei, J., Zhang, X., Sheng, Y., Shen, J., Huang, P., Guo, S., Pan, J., Liu, B., and Feng, B. Simple one-step synthesis of water-soluble fluorescent carbon dots from waste paper. New Journal of Chemistry, 2014, 38(3), 906.

[81] Pandey, S., Mewada, A., Thakur, M., Pillai, S., Dharmatti, R., Phadke, C., and Sharon, M. Synthesis of mesoporous silica oxide/C-dot complex (meso-SiO_2/C-dots) using pyrolysed rice husk and its application in bioimaging. RSC Advances, 2014, 4(3), 1174.

[82] Zhao, X. J., Zhang, W. L., and Zhou, Z. Q. Sodium hydroxide-mediated hydrogel of citrus pectin for preparation of fluorescent carbon dots for bioimaging. Colloids and Surfaces B: Biointerfaces, 2014, 123, 493.

[83] Wang, Q., Zhang, S., Ge, H., Tian, G., Cao, N., and Li, Y. A fluorescent turn-off/on method based on carbon dots as fluorescent probes for the sensitive determination of Pb^{2+} and pyrophosphate in an aqueous solution. Sensors and Actuators B: Chemical, 2015, 207, Part A, 25.

[84] Barati, A., Shamsipur, M., Arkan, E., Hosseinzadeh, L., and Abdollahi, H. Synthesis of biocompatible and highly photoluminescent nitrogen doped carbon dots from lime: Analytical applications and optimization using response surface. Materials Science and Engineering: C, 2015, 47, 325.

[85] Xu, Y., Tang, C.-J., Huang, H., Sun, C.-Q., Zhang, Y.-K., Ye, Q.-F., and Wang, A.-J. Green synthesis of fluorescent carbon quantum dots for detection of Hg^{2+}. Chinese Journal of Analytical Chemistry, 2014, 42(9), 1252.

[86] Zhang, J., Yuan, Y., Liang, G., and Yu, S.-H. Scale-Up Synthesis of Fragrant Nitrogen-Doped Carbon Dots from Bee Pollens for Bioimaging and Catalysis. Advanced Science, 2015, 2(4).

[87] Liu, R., Zhang, J., Gao, M., Li, Z., Chen, J., Wu, D., and Liu, P. A facile microwave-hydrothermal approach towards highly photoluminescent carbon dots from goose feathers. RSC Advances, 2015, 5(6), 4428.

[88] Mehta, V. N., Jha, S., and Kailasa, S. K. One-pot green synthesis of carbon dots by using Saccharum officinarum juice for fluorescent imaging of bacteria (Escherichia coli) and yeast (Saccharomyces cerevisiae) cells. Materials Science and Engineering: C, 2014, 38, 20.

[89] Wolfbeis, O. S. An overview of nanoparticles commonly used in fluorescent bioimaging. Chemical Society Reviews, 2015, 44(14), 4743.

[90] Hola, K., Zhang, Y., Wang, Y., Giannelis, E. P., Zboril, R., and Rogach, A. L. Carbon dots – Emerging light emitters for bioimaging, cancer therapy and optoelectronics. Nano Today, 2014, 9(5), 590.

[91] Zheng, X. T., Ananthanarayanan, A., Luo, K. Q., and Chen, P. Glowing graphene quantum dots and carbon dots: properties, syntheses, and biological applications. Small, 2015, 11(14), 1620.

[92] Li, Y.-Q., Samad, Y. A., Polychronopoulou, K., and Liao, K. Lightweight and highly conductive aerogel-like carbon from sugarcane with superior mechanical and EMI shielding properties. ACS Sustainable Chemistry & Engineering, 2015, 3(7), 1419.

[93] Sun, H., Ji, H., Ju, E., Guan, Y., Ren, J., and Qu, X. Synthesis of Fluorinated and Nonfluorinated Graphene Quantum Dots through a New Top-Down Strategy for Long-Time Cellular Imaging. Chemistry – A European Journal, 2015, 21(9), 3791.

[94] Mohan, D., Singh, K. P., and Singh, V. K. Wastewater treatment using low cost activated carbons derived from agricultural byproducts – a case study. Journal of Hazardous Materials, 2008, 152(3), 1045.

[95] Bhatnagar, A., and Sillanpää, M. Utilization of agro-industrial and municipal waste materials as potential adsorbents for water treatment – a review. Chemical Engineering Journal, 2010, 157(2–3), 277.

[96] Ayotamuno, M. J., Kogbara, R. B., Ogaji, S. O. T., and Probert, S. D. Petroleum contaminated ground-water: Remediation using activated carbon. Applied Energy, 2006, 83(11), 1258.
[97] Unur, E. Functional nanoporous carbons from hydrothermally treated biomass for environmental purification. Microporous and Mesoporous Materials, 2013, 168(0), 92.
[98] Qu, X., Brame, J., Li, Q., and Alvarez, P. J. J. Nanotechnology for a safe and sustainable water supply: enabling integrated water treatment and reuse. Accounts of Chemical Research, 2013, 46(3), 834.
[99] Li, H., Gui, X., Zhang, L., Wang, S., Ji, C., Wei, J., Wang, K., Zhu, H., Wu, D., and Cao, A. Carbon nanotube sponge filters for trapping nanoparticles and dye molecules from water. Chemical Communications, 2010, 46(42), 7966.
[100] de Vries, W., Römkens, P. A. M., and Schütze, G. Reviews of environmental contamination and toxicology. In: *Critical Soil Concentrations of Cadmium, Lead, and Mercury in View of Health Effects on Humans and Animals*. Springer New York, 2007, 191, 91–130.
[101] Nolan, E. M., and Lippard, S. J. Tools and tactics for the optical detection of mercuric ion. Chemical Reviews, 2008, 108(9), 3443.
[102] Xu, Z., Han, S. J., Lee, C., Yoon, J., and Spring, D. R. Development of off-on fluorescent probes for heavy and transition metal ions. Chemical Communications, 2010, 46(10), 1679.
[103] Wang, X., Liu, J., and Xu, W. One-step hydrothermal preparation of amino-functionalized carbon spheres at low temperature and their enhanced adsorption performance towards Cr(VI) for water purification. Colloids and Surfaces A: Physicochemical and Engineering Aspects, 2012, 415, 288.
[104] Åkesson, A., Julin, B., and Wolk, A. Long-term dietary cadmium intake and postmenopausal endometrial cancer incidence: a population-based prospective cohort study. Cancer Research, 2008, 68(15), 6435.
[105] Liu, Y., Zhao, Y., and Zhang, Y. One-step green synthesized fluorescent carbon nanodots from bamboo leaves for copper(II) ion detection. Sensors & Actuators: B. Chemical, 2014, 196 (Complete), 647.
[106] Fu, Y., Ding, C., Zhu, A., Deng, Z., Tian, Y., and Jin, M. Two-photon ratiometric fluorescent sensor based on specific biomolecular recognition for selective and sensitive detection of copper ions in live cells. Analytical Chemistry, 2013, 85(24), 11936.
[107] Godbold, D. L., Fritz, E., and Hüttermann, A. Aluminum toxicity and forest decline. Proceedings of the National Academy of Sciences of the United States of America, 1988, 85(11), 3888.
[108] Goswami, S., Paul, S., and Manna, A. Selective "naked eye" detection of Al (III) and PPi in aqueous media on a rhodamine-isatin hybrid moiety. RSC Advances, 2013, 3(27), 10639.
[109] Yang, S.-T., Wang, X., Wang, H., Lu, F., Luo, P. G., Cao, L., Meziani, M. J., Liu, J.-H., Liu, Y., Chen, M., Huang, Y., and Sun, Y.-P. Carbon dots as nontoxic and high-performance fluorescence imaging agents. The Journal of Physical Chemistry C, 2009, 113(42), 18110.
[110] Shi, L., Li, X., Li, Y., Wen, X., Li, J., Choi, M. M. F., Dong, C., and Shuang, S. Naked oats-derived dual-emission carbon nanodots for ratiometric sensing and cellular imaging. Sensors and Actuators B: Chemical, 2015, 210, 533.
[111] Lu, W., Qin, X., Liu, S., Chang, G., Zhang, Y., Luo, Y., Asiri, A. M., Al-Youbi, A. O., and Sun, X. Economical, green synthesis of fluorescent carbon nanoparticles and their use as probes for sensitive and selective detection of mercury (II) ions. Analytical Chemistry, 2012, 84(12), 5351.
[112] Tan, X. W., Romainor, A. N. B., Chin, S. F., and Ng, S. M. Carbon dots production via pyrolysis of sago waste as potential probe for metal ions sensing. Journal of Analytical and Applied Pyrolysis, 2014, 105, 157.
[113] Suryawanshi, A., Biswal, M., Mhamane, D., Gokhale, R., Patil, S., Guin, D., and Ogale, S. Large scale synthesis of graphene quantum dots (GQDs) from waste biomass and their use as an

efficient and selective photoluminescence on -off -on probe for Ag^+ ions. Nanoscale, 2014, 6(20), 11664.
[114] Yeung, P.-T., Chung, P.-Y., Tsang, H.-C., Cheuk-On Tang, J., Yin-Ming Cheng, G., Gambari, R., Chui, C.-H., and Lam, K.-H. Preparation and characterization of bio-safe activated charcoal derived from coffee waste residue and its application for removal of lead and copper ions. RSC Advances, 2014, 4(73), 38839.
[115] Tang, J., Mu, B., Zheng, M., and Wang, A. One-step calcination of the spent bleaching earth for the efficient removal of heavy metal ions. ACS Sustainable Chemistry & Engineering, 2015, 3(6), 1125.
[116] Hu, Y., Yang, J., Tian, J., Jia, L., and Yu, J.-S. Green and size-controllable synthesis of photoluminescent carbon nanoparticles from waste plastic bags. RSC Advances, 2014, 4(88), 47169.
[117] Li, T., Shen, J., Huang, S., Li, N., and Ye, M. Hydrothermal carbonization synthesis of a novel montmorillonite supported carbon nanosphere adsorbent for removal of Cr (VI) from waste water. Applied Clay Science, 2014, 93–94, 48.
[118] Konicki, W., Cendrowski, K., Bazarko, G., and Mijowska, E. Study on efficient removal of anionic, cationic and nonionic dyes from aqueous solutions by means of mesoporous carbon nanospheres with empty cavity. Chemical Engineering Research and Design, 2015, 94, 242.
[119] Li, M., Wang, C., O'Connell, M. J., and Chan, C. K. Carbon nanosphere adsorbents for removal of arsenate and selenate from water. Environmental Science: Nano, 2015, 2(3), 245.
[120] Song, X., Gunawan, P., Jiang, R., Leong, S. S. J., Wang, K., and Xu, R. Surface activated carbon nanospheres for fast adsorption of silver ions from aqueous solutions. Journal of Hazardous Materials, 2011, 194, 162.
[121] Yang, S. J., Kang, J. H., Jung, H., Kim, T., and Park, C. R. Preparation of a freestanding, macroporous reduced graphene oxide film as an efficient and recyclable sorbent for oils and organic solvents. Journal of Materials Chemistry A, 2013, 1(33), 9427.
[122] Peterson, C. H., Rice, S. D., Short, J. W., Esler, D., Bodkin, J. L., Ballachey, B. E., and Irons, D. B. Long-term ecosystem response to the Exxon Valdez oil spill. Science, 2003, 302(5653), 2082.
[123] Cheng, M., Gao, Y., Guo, X., Shi, Z., Chen, J.-f., and Shi, F. A functionally integrated device for effective and facile oil spill cleanup. Langmuir, 2011, 27(12), 7371.
[124] Carrier, G., Fendell, F., and Mitchell, J. In situ burning via towed boom of oil spilled at sea. Combustion and Flame, 1992, 90(3–4), 295.
[125] Konishi, M., Kishimoto, M., Tamesui, N., Omasa, T., Shioya, S., and Ohtake, H. The separation of oil from an oil – water – bacteria mixture using a hydrophobic tubular membrane. Biochemical Engineering Journal, 2005, 24(1), 49.
[126] Bi, H., Yin, Z., Cao, X., Xie, X., Tan, C., Huang, X., Chen, B., Chen, F., Yang, Q., Bu, X., Lu, X., Sun, L., and Zhang, H. Carbon fiber aerogel made from raw cotton: a novel, efficient and recyclable sorbent for oils and organic solvents. Advanced Materials, 2013, 25(41), 5916.
[127] Hashim, D. P., Narayanan, N. T., Romo-Herrera, J. M., Cullen, D. A., Hahm, M. G., Lezzi, P., Suttle, J. R., Kelkhoff, D., Muñoz-Sandoval, E., Ganguli, S., Roy, A. K., Smith, D. J., Vajtai, R., Sumpter, B. G., Meunier, V., Terrones, H., Terrones, M., and Ajayan, P. M. Covalently bonded three-dimensional carbon nanotube solids via boron induced nanojunctions. Scientific Reports, 2012, 2(363), 1.
[128] Zhang, S., Zeng, M., Li, J., Li, J., Xu, J., and Wang, X. Porous magnetic carbon sheets from biomass as an adsorbent for the fast removal of organic pollutants from aqueous solution. Journal of Materials Chemistry A, 2014, 2(12), 4391.

[129] Mittal, A., Malviya, A., Kaur, D., Mittal, J., and Kurup, L. Studies on the adsorption kinetics and isotherms for the removal and recovery of Methyl Orange from wastewaters using waste materials. Journal of Hazardous Materials, 2007, 148(1–2), 229.
[130] Chatterjee, D., and Dasgupta, S. Visible light induced photocatalytic degradation of organic pollutants. Journal of Photochemistry and Photobiology C: Photochemistry Reviews, 2005, 6(2–3), 186.
[131] Crini, G. Non-conventional low-cost adsorbents for dye removal: a review. Bioresource Technology, 2006, 97(9), 1061.
[132] Myung, Y., Jung, S., Tung, T. T., Tripathi, K. M. and Kim, T-Y. Graphene-based aerogels derived from biomass for energy storage and environmental remediation. ACS Sustainable Chemistry & Engineering, 2019, 7, 2772.
[133] Chandra, V., Park, J., Chun, Y., Lee, J. W., Hwang, I.-C., and Kim, K. S. Water-dispersible magnetite-reduced graphene oxide composites for arsenic removal. ACS Nano, 2010, 4(7), 3979.
[134] Sun, R., Wang, Y., Ni, Y., and Kokot, S. Graphene quantum dots and the resonance light scattering technique for trace analysis of phenol in different water samples. Talanta, 2014, 125, 341.
[135] Magnacca, G., Allera, A., Montoneri, E., Celi, L., Benito, D. E., Gagliardi, L. G., Gonzalez, M. C., Mártire, D. O., and Carlos, L. Novel magnetite nanoparticles coated with waste-sourced biobased substances as sustainable and renewable adsorbing materials. ACS Sustainable Chemistry & Engineering, 2014, 2(6), 1518.
[136] Mahanta, N., Leong, W. Y., and Valiyaveettil, S. Isolation and characterization of cellulose-based nanofibers for nanoparticle extraction from an aqueous environment. Journal of Materials Chemistry, 2012, 22(5), 1985.
[137] Shukla, S., Khan, I., Bajpai, V. K., Lee, H., Kim, T. Y., Upadhyay, A., Huh, Y. S., Han, Y. K., and Tripathi, K. M. Sustainable graphene aerogel as an ecofriendly cell growth promoter and highly efficient adsorbent for histamine from red wine. ACS Applied Materials & Interphases, 2019, 11, 18165.
[138] Liu, S., Wei, L., Hao, L., Fang, N., Chang, M. W., Xu, R., Yang, Y., and Chen, Y. Sharper and faster "nano darts" kill more bacteria: a study of antibacterial activity of individually dispersed pristine single-walled carbon nanotube. ACS Nano, 2009, 3(12), 3891.
[139] Liu, S., Zeng, T. H., Hofmann, M., Burcombe, E., Wei, J., Jiang, R., Kong, J., and Chen, Y. Antibacterial Activity of Graphite, Graphite Oxide, Graphene Oxide, and Reduced Graphene Oxide: Membrane and Oxidative Stress. ACS Nano, 2011, 5(9), 6971.
[140] Brady-Estévez, A. S., Kang, S., and Elimelech, M. A single-walled-carbon-nanotube filter for removal of viral and bacterial pathogens. Small, 2008, 4(4), 481.
[141] Chen, J., Peng, H., Wang, X., Shao, F., Yuan, Z., and Han, H. Graphene oxide exhibits broad-spectrum antimicrobial activity against bacterial phytopathogens and fungal conidia by intertwining and membrane perturbation. Nanoscale, 2014, 6(3), 1879.
[142] Mohanty, B., Anita, K. V., Claesson, P., and Bohidar, H. B. Physical and anti-microbial characteristics of carbon nanoparticles prepared from lamp soot. Nanotechnology, 2007, 18(44), 445102.
[143] Varghese, S., Kuriakose, S., and Jose, S. Antimicrobial activity of carbon nanoparticles isolated from natural sources against pathogenic Gram-negative and Gram-positive bacteria. Journal of Nanoscience, 2013, 2013, 5.
[144] Das Purkayastha, M., Das, S., Manhar, A. K., Deka, D., Mandal, M., and Mahanta, C. L. Removing Antinutrients from Rapeseed Press-Cake and Their Benevolent Role in Waste Cooking Oil-Derived Biodiesel: Conjoining the Valorization of Two Disparate Industrial Wastes. Journal of Agricultural and Food Chemistry, 2013, 61(45), 10746.

Seema Garg, Rohit Bhatia, and Pankaj Attri

3 Black but gold: carbon nanomaterials for waste water purification

3.1 Introduction

With the advent of twenty-first century, remarkable milestones in terms of urbanization, industrialization, and digitalization in all sectors of livelihood have been laid. They, on the one hand, undoubtedly exalted the living standards of people worldwide but at the same time imposed grave issues at the basic humanitarian ground too. Inadequate access to clean, safe, and affordable water arises as the most dreadful implication of this development as water is the fundamental necessity for all forms of life on earth. Incessant dumping of domestic, commercial, industrial, and agricultural effluents (the concealed villains of urbanization and industrialization) in natural water resources is gradually restraining the fresh water supply together with severely hampering the aquatic life. The situation is further exacerbated by unusual global climate change and poor water use practices and wastewater management strategies [1]. The consequential environmental condition not only disturbs the social network and human well-being but also jeopardizes economic growth of the nation's worldwide. According to World Health Organization and United Nations International Children's Emergency Fund, one-third of the total world's population still lack reliable access to clean water for drinking and basic sanitation. This counts for almost 3.1% of death worldwide, becoming the second largest cause for the death of children [2].

Since growth and development is indispensable, the better alternative (to combat scarcity of fresh, clean and safe water) lies in ensuring proficient wastewater management practices rather than deterring the technorevolution. Wastewater can be broadly described as a combination of various pollutants and contaminants including plant nutrients (nitrogen, phosphorus, potassium), pathogenic microorganisms (viruses, bacteria, protozoa, and helminths), heavy metals (e.g., cadmium, chromium, copper, mercury, nickel, lead, and zinc), organic pollutants (e.g., polychlorinated biphenyls, polyaromatic hydrocarbons, pesticides), and micropollutants (e.g., medicines, cosmetics, cleaning agents). Management of such water refers to its sustainable treatment from source to re-entry into the environment such that its detrimental effects are minimized. UN water statistics indicate that presently only 20% of globally produced wastewater receives proper treatment [3].

Seema Garg, Department Of Chemistry, University Of Delhi, Delhi, India
Rohit Bhatia, Department Of Chemistry, Institute Of Home Economics, Delhi, India; J&S Research and Innovations, New Delhi, India
Pankaj Attri, Plasma Bioscience Research Center/ Department of Electrical and Biological Physics, Kwangwoon University, Seoul, Republic of Korea

https://doi.org/10.1515/9783110650600-003

Arrays of aerobic, anaerobic, and physicochemical processes that can treat wastewaters to almost any standard of effluent (from the simple removal of gross solids to membrane systems) have been discovered time and again (Table 3.1). They vary from very simple to highly complex and each has its own characteristics in terms of efficiency, reliability, cost, affordability, energy consumption, and so on [3]. Most of these methods, however, rely on conveyance and centralized systems for their operation and, therefore, are no longer sustainable [4]. In addition, factors like diverse influent water quality, novel toxic compounds of anthropogenic origin, lethal by-products during purification, and harmful toxins (released by microorganisms in the course of treatment) further restrict the flexibility of these conventional wastewater treatment technologies [5]. Therefore, a paradigm shift in reframing the water treatment technologies, essentially from point of entry (or off-site) approaches to point of use (POU or on-site) technologies (exhibiting multifaceted and modular applications) so as to establish next-generation water purification systems is required.

In this context, nanotechnology offers bourgeoning opportunities owing to the distinctive structural, chemical, and physical characteristics of materials at this scale. A plethora of techniques involving nanostructured materials (namely, nano-TiO_2, nanoiron, metallic nanoparticles and their oxides, nanozeolites, quantum dots, nanosilica, etc.) congregated with the current water treatment approaches have been introduced in recent times [6]. Some of these techniques simply utilize the smoothly scalable size-dependent properties of nanomaterials (which relate to the high specific surface area (SSA)) such as fast dissolution, high reactivity, and strong sorption, while others take advantage of their discontinuous properties, such as superparamagnetism, localized surface plasmon resonance, and quantum confinement effect [4]. These techniques undoubtedly surpassed the traditional strategies in terms of efficiency, reliability, cost, time, reuse, and versatility. However despite several advantages, they also suffer from some serious disadvantages such as ineffectiveness under sunlight, thermal instability, concentration problems, formation of more toxic intermediates, fouling and precipitation of pollutants, pore blocking, and so on [6].

Carbon nanomaterials, by far, have shown supreme potential in overcoming the major challenges posed by both the raw/traditional as well as nanoenabled water purification systems. The triumph of carbon nanomaterials is an accomplishment of both their unique structural, mechanical, chemical, and electrical properties together with their incomparable thermal stability, flexibility, and suitability for organic, inorganic, and biological water pollutants [5]. These nanomaterials chiefly encompass carbon nanotubes (CNTs), fullerenes, graphene, and their derivatives with CNTs being the most studied and explored member. They have been frequently utilized in various water purification systems as nanosorbants, nanofilters, nanomembranes, disinfectants, nanosensors, or nanocatalysts. In the next section, the role of different carbon nanomaterials in various wastewater treatment technologies along with the benefits offered by them over conventional counterparts has been highlighted.

Table 3.1: Potential of CNTs in removal of different water pollutants via adsorption.

Adsorbant	Adsorbate	Comments	Ref.
		Removal of organic pollutants	
As-grown CNTs and graphitized CNTs	1,2-Dichlorobenzene (DCB)	As grown CNTs showed better adsorption than graphitized CNTs due to smooth surface of the latter. Removal efficiency of DCB by both as-grown CNTs and graphitized CNTs remained stable in the pH range 3–10.	[19]
CNTs purified by mixed HNO_3 and H_2SO_4	Polycyclic aromatic hydrocarbons (PAHs)	SWCNTs exhibited a larger adsorption capacity toward PAHs than MWCNTs.	[27]
CNTs purified by mixed HNO_3 and H_2SO_4	Phenol, pyrogallol, and 1-naphthol	CNTs with smaller outer diameter had higher distribution coefficients (K_d). The K_d values of the three polar phenolics tended to increase with an increase in pH and then decrease with pH over their pK_a value.	[16]
CNTs purified by mixed HNO_3 and H_2SO_4	Triton X-series surfactants	Hydrophobic and π–π interactions between the surfactants and CNTs were the dominant mechanisms. The adsorption of Triton X-series surfactants facilitated suspending CNTs in water.	[45]
CNTs treated with HCL and HNO_3	Basic orange and methyl violet (6B)	The removal percentage of organic dyes increased with an increase in the contact time, CNTs dosage, and pH and decreased with an increase in concentrations of dyes. The maximum removal percentage for all dyes concentrations was at pH = 8.5 and CNTs dosage of 0.25 g/L and 0.3 g/L for methyl violet and basic orange dye, respectively.	[46]
CNTs activated by KOH etching	Sulfamethoxazole, tetracycline, and tylosin	KOH etching improved the adsorption affinity and adsorption reversibility of pharmaceutical antibiotics.	[47]

As-prepared and oxidized MWCNTs	Ionizable aromatic Compounds (IACs): 1-naphthylamine, 1-naphthol, phenol	The adsorption capacity of MWCNTs toward IACs was higher than other common adsorbents such as natural bentonite, apatite, and kaolinite. Oxidation of MWCNTs increased the surface area and added oxygen-containing functional groups to the surfaces of MWCNTs, which depressed the adsorption of IACs on MWCNTs.	[48]
As-prepared MWCNT	Methylene blue (MB) and Acid Red 183 (AR183)	MWCNT presented the maximum adsorption capacities of MB and AR183 at 59.7 and 45.2 mg/g, respectively. MWCNTs showed higher adsorption of MB than AR183 in both single and binary dye systems. π–π stacking was the main driving force responsible for the dye–MWCNT interaction.	[49]
As-prepared MWCNTs	Amoxicillin	MWCNTs (0.1 g and 0.2 g) showed 86.5% and 87.9% adsorption efficiency in continuous assays. Adsorption rate was much higher than fullerene C_{60} and AC. Adsorption capacity of 22.9 mg amoxicillin per gram of MWCNT was achieved.	[50]
As-prepared MWCNTs	Rhodamine B	Highest adsorption capacity of 65–90% for the dye by MWCNT was achieved. Increased concentration of Rhodamine B demonstrated higher adsorption %. Enhanced incubation time provided maximum adsorption of Rhodamine B.	[51]
As-prepared MWCNTs	Methylene blue (MB) and Congo red	Adsorption isotherms indicated an absorbance of 400 mg/g and 500 mg/g for MB and Congo red, respectively. Increase in the contact time resulted in greater adsorption % Adsorption rate was independent of dye concentration and solution temperature.	[52]
TiO_2/graphene MWCNTs nanocomposite	Methylene blue	Presence of TiO_2 decreased the overall adsorption capacity by blocking the adsorption sites of CNTs.	[53]

(continued)

Table 3.1 (continued)

Adsorbant	Adsorbate	Comments	Ref.
Amine functionalized magnetic CNTs	Acid Blue 92 (AB92), Acid Red 14 (AR14), and Direct Red 31 (DR31)	Highest adsorption capacity for AB192, AR14, and DR31 was 333, 370, and 323 mg/g, respectively. Amine substituent aided in the effective removal of cationic dyes.	[54]
MWCNT–polyacrylonitrile electrospun nanofibers (PEN)	Pyrene	Doped unzipped MWCNT (2%) into the composite nanofiber achieved 96% pyrene adsorption in 30 min at RT. Recycling of the nanocomposite up to 6 cycles was observed without significant loss in efficiency and morphology.	[55]
As-prepared MWCNTs	Direct Blue 53 (DB53)	Maximum sorption capacity occurred at 323 K, with the values of 409.4 and 135.2 mg/g for MWCNT and AC, respectively. DB-53 loaded MWCNT could be regenerated (97.85%) using a mixture 50% acetone + 50% of 3 mol L^{-1} NaOH. Optimum adsorption of the dye was achieved at pH 2.0.	[56]
		Removal of heavy metal ions	
MWCNTs	Pb^{2+}	Maximum Pb^{2+} adsorption occurred at pH5.0 and high temperature.	[57]
As-produced MWCNTs	Ni^{2+}	The nickel ions uptake for the as-produced CNTs increased slowly and reached the maximum uptake of 12.3 mg/g after 50 minutes for initial nickel ions concentration of 200 mg/L.	[58]
MWCNTs/iron Oxides	Ni^{2+} Sr^{2+}	The adsorption capacity of the composite for Sr^{2+} is higher than that of MWCNTs and iron oxides at pH > 7. Adsorption capacity for Ni^2 was 9.18 mg/g, whereas removal % was 87% (Ni^{2+}) and 97 %(Sr^{2+}) at pH 6.5 with CNT dosage of 2 g at 298 K.	[59]

As-grown CNTs	Cd^{2+}	Maximum adsorption of 3.5 mg/g was achieved at the high CNT dosage of 0.2 g per 100 ml.	[60]
Oxidized MWCNTs	Cu^{2+}, Zn^{2+}, Pb^{2+}, Cd^{2+}, and Co^{2+}	The order of adsorption onto the oxidized CNT sheets was $Pb^{2+} > Cd^{2+} > Co^{2+} > Zn^{2+} > Cu^{2+}$. The adsorption capacity (mg/g) of the various ions was 50.37 (Cu^{2+}), 58 (Zn^{2+}), 101.05 (Pb^{2+}), 75.84 (Cd^{2+}), 69.63 (Co^{2+}) with removal % of 41 (Cu^{2+}), 66 (Zn^{2+}), 84 (Pb^{2+}), 63 (Cd^{2+}), and 58 (Co^{2+}), respectively.	[61]
Oxidized MWCNTs and ethylenediamine functionalized MWCNTs (e-MWCNT)	Cd^{2+}	Enhanced adsorption efficiency of 25.7 mg/g was achieved by e-MWCNTs in comparison to 23.32 mg/g for oxidized MWCNTs.	[62]
As-prepared, HNO_3-modified, and NaOCl-modified MWCNTs	Cu^{2+}	Negatively charged surface of modified CNTs increased the adsorption rate and adsorption sites. The adsorption capacity of Cu^{2+} followed the order NaOCl-modified CNTs (47.39 mg/g) > HNO_3-modified CNTs (13.87 mg/g) > as-produced CNTs.	[63]
MWCNTs and SWCTNs	Ni^{2+} Zn^{2+}	Sorption affinity between Zn^{2+} and CNT surface (14.4 mg/g) was stronger than that between Ni^{2+} and CNT surface (13.05 mg/g).	[64]
MWCNTs-activated carbon (AC)	Cr^{6+}	Excellent adsorption capacity of 9.0 mg/g at pH 2–4, contact time 60 min, agitation speed 100–200 rpm was achieved.	[65]
Amorphous Al_2O^3 supported on CNTs	F^-	The adsorption performed well at pH 5–9, which was a much broader range than that of the activated alumina (pH < 6).	[66]
Ceria nanoparticles supported on CNTs	As^{5+}	The As(V)-loaded adsorbent could be efficiently regenerated. Ca^{2+} and Mg^{2+} ions in water enhanced the adsorption capacity of CeO_2-CNTs toward arsenate due to the formation of ternary surface complex.	[67]
3-Mercaptopropyltrimethoxysilane (MPTMS) functionalized MWCNTs	Pb^{2+}	3-MPTMS-functionalized MWCNTs improved the sensitivity for the Pb^{2+} trace determination to 95% compared with the oxidized MWCNTs.	[68]

(continued)

Table 3.1 (continued)

Adsorbant	Adsorbate	Comments	Ref.
Diglycolamide (DGA) functionalized MWCNTs	Uranium	DGA-MWCNTs showed uranium adsorption from aqueous solution with K_d value as high as 4×10^4 mL g^{-1} at 4 M HNO3 and at room temperature.	[69]
$HNO_3 \bullet KMnO_4$ and H_2SO_4-oxidized MWCNTs	Tl(III)	MWCNTs oxidized by HNO_3 showed the best sorption performance for Tl (III) ions. Maximum adsorption of Tl(III) was recorded at a pH value of 7. Maximum amounts of Tl(III) adsorbed on raw MWCNTs, H_2SO_4-oxidized, $KMnO_4$-oxidized, and HNO_3-oxidized MWCNTs as calculated by the Langmuir model were 3.0, 11.7, 21.6, and 31.5 mg/g, respectively.	[70]
Tris(2-aminoethyl)amine [TAA] grafted MWCNTs	Pb^{2+}	MWCNTs-TAA offered much better selectivity for Pb(II) than other heavy metals. Maximum adsorption capacity of 38 mg/g was achieved at pH above 4.	[71]
		Removal of biological pollutants	
Ag-MWCNTs/β-cyclo dextrin	*Escherichia coli*	The composite showed antibacterial activity via bacterial membrane disruption and *E. coli* growth delay.	[72]
Polyurethane sponge coated with CNTs and silver nanowires	Bacteria (E.coli, S. enterica Typhimirium, E.faecalis, and B.subtilis); Virus (bacteriophage MS2)	The composite achieved > 6 log (99.9999%) removal of four model bacteria, including *E. coli*, *S. enterica* Typhimirium, *E. faecalis*, and *B. subtilis*. Also showed > 2 log (99%) removal of bacteriophage MS2, with a low energy consumption of only 100 J/L.	[73]

Pristine (P-)MWCNTs **Oxidized (O-)MWCNTs** **ZnO/MWCNTs**	*E. coli*	ZnO/MWCNTs showed strongest antimicrobial activity compared with P-MWCNTs and O-MWCNTs. ZnO/MWCNTs (0.128 mg mL^{-1}), indicated minimum inhibition zone (MIZ) of *E. coli*, i.e., 0.128 mg mL^{-1}. Deposited ZnO was important for bactericidal action of ZnO/ MWCNTs, while the P-MWCNTs and O-MWCNTs served as more like adsorbing materials for *E. coli*.	[74]
SWCNTs	*Bacillus anthracis* (*B. anthracis*)	Combined treatment with SWCNTs (100 µg/mL) and H_2O_2 (1.5%) or NaOCl (0.25%) exhibited much stronger sporicidal effect on the spores, compared to treatment with H_2O_2 or NaOCl alone.	[75]
Cu–polyacrylic acid (PAAc)/MWCNTs	*E. coli*	Cu ions played crucial role in the antimicrobial activity of the nanocomposite.	[76]
Pristine MWCNTs **MWCNTs–arginine** **MWCNTs–Lysine**	*E. coli*, *S. Typhimurium*, and S. *aureus*	Antibacterial activity trend followed the order MWCNTs–arginine > MWCNTs–Lysine > pristine MWCNTs. Pristine MWCNT samples were effective against the resistant strain *S. aureus*. Grafted MWCNT samples were highly effective against *E. coli* and *S. Typhimurium*.	[77]

3.2 CNTs in wastewater purification

CNTs as cited by Tan are seamless cylinder-shaped macromolecules with a radius as small as a few nanometers and up to several micrometers in length. Their walls are composed of graphite sheets (allotropic form of carbon) rolled up into a tubelike structure with the appearance of latticework fence. Depending upon the number of graphene sheet layers, they can be single-walled carbon nanotubes (SWCNTs) or multiwalled carbon nanotubes (MWCNTs). The latter are composed of two or more concentric cylindrical shells of graphene sheets coaxially arranged around a central hollow area, whereas SWCNTs are made of a single cylinder graphite sheet held together by van der Waals bonds [7]. The high SSA, tunable surface reactivity, and hollow-layered structure of CNTs make them one of the most ideal candidates for applications in the wastewater treatment strategies. Owing to their distinguished characteristics, CNTs have found copious applications in water purification processes as sorbents, catalyst, filters or membranes, disinfectants, and so on (Figure 3.1).

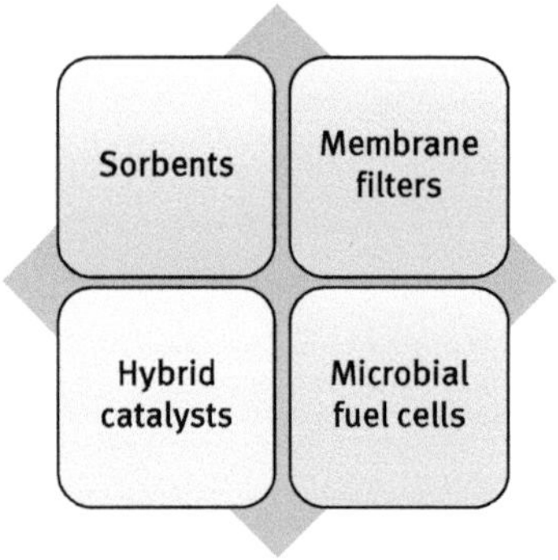

Figure 3.1: Role of CNTs in wastewater treatment processes.

3.3 CNTs as nanosorbants

Adsorption has always been considered as the key mechanism for the removal of variety of pollutants (organic, inorganic, and biological) from wastewater. Basically, it is a surface phenomenon that involves the adsorption of any pollutant (referred as adsorbate) on the surface of adsorbent, thereby forming a thin film over it [8]. The extent and strength of adsorption is largely a function of the type of adsorbent. Noxious pollutants such as pesticides, heavy metals, dyes, and other organic molecules have always spawned the need of effective adsorbents that not only exhibit multiple adsorption sites with high specific area but also display fast adsorption kinetics. Traditionally, adsorbents such as activated carbons (ACs), zeolites, and resins were widely used; however, slow adsorption kinetics, lack of selectivity, and difficulty for regeneration constrained their applicability with time. The era of CNTs marked the beginning of a new generation adsorbents in the field of wastewater purification. They, besides holding on to the advantages offered by conventional adsorbents (such as, large surface area, thermal stability, chemical inertness, and porous structure), also promise to overcome their major shortcomings. CNTs, as reported earlier, are long, fibrous uniform cylindrical-shaped nanostructures with high aspect ratio. Their splendid adsorption capacity is attributable to various factors that have been highlighted here.

3.3.1 Exceptionally high specific surface area with associated adsorption sites

The one-dimensional fibrous nanostructure of CNTs offers large surface-to-volume ratio, which calls for their exceptionally high SSA. The adsorption sites in the individual/nonaggregated CNTs are essentially the open-ended portions and their external surface [9]. However in aqueous solution, CNTs are prone to form loose bundles/aggregates due to the hydrophobicity of their graphitic surface. Aggregation reduces the effective surface area of CNTs, thereby lowering their adsorption capacity, but at the same time it also promotes the formation of some distinguished sites like interstitial spaces and grooves (Figure 3.2), which significantly increase the pore volume

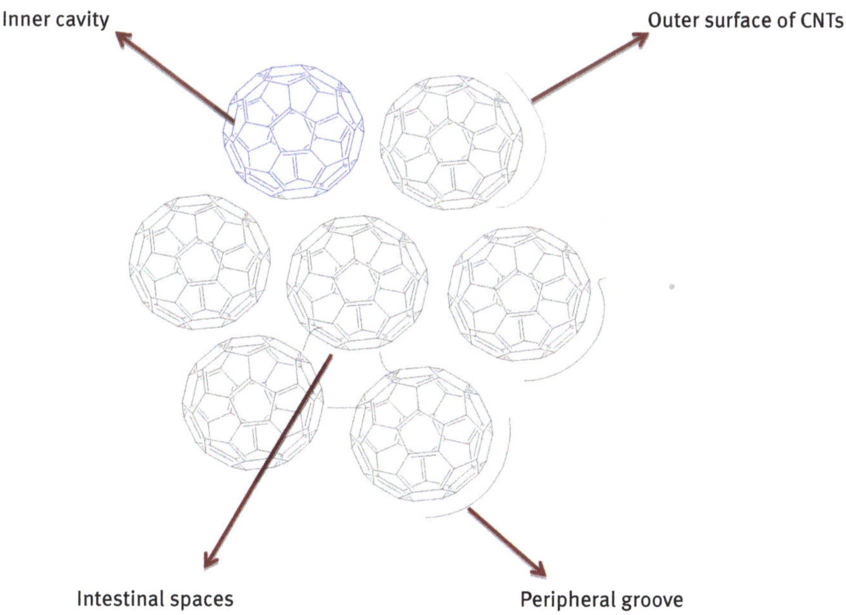

Figure 3.2: A pictorial illustration of various adsorption sites in aggregated CNTs.

and act as high adsorption energy sites for various other bulky organic pollutants (such as pharmaceuticals, cosmetics, etc.) that are otherwise not adsorbed by conventional adsorbents [10, 11].

3.3.2 Fast adsorption kinetics

For a solid–liquid sorption system essentially four steps constitute the adsorption process, namely, bulk diffusion, boundary layer diffusion (external mass transfer), intraparticle diffusion, and surface reaction. Bulk diffusion is the transport of solute from the bulk solution to the boundary layer surrounding the adsorbent particles, and diffusion through the boundary layer to the sorbent exterior surface is referred as *boundary layer diffusion* (external mass transfer). Intraparticle diffusion is the migration of solute within the pores of the adsorbent along the sorbent inner surface (surface diffusion) and the surface reaction is the interaction of solute with the available sites on the interior surface of the pores and capillary spaces of the sorbent. The overall rate of adsorption is controlled by the slowest step. Generally, the bulk diffusion and surface reaction steps are rapid and not rate-limiting [12]. The uniform, homogeneous, and porous structure of CNTs allows rapid intraparticle diffusion of the adsorbed molecule due to short intraparticle distance, thereby reducing the time taken to reach the equilibrium and hence displays fast adsorption kinetics. The kinetics has been shown to be further accentuated by easy and highly accessible adsorption sites on the external surface of CNTs [12,13].

3.3.3 Diverse types of interaction with pollutant

One of the most distinguished characteristic of CNTs that explains their brilliant adsorption capacity is their ability to interact with pollutants via diverse types of interactions including physical, covalent, noncovalent, and electrostatic interactions [9]. These interactions either operate singly or in conjunction with each other. Archetypally, the surface of pristine (unmodified) CNTs is hydrophobic in nature owing to the presence of nonpolar graphene sheets. The hydrophobicity grants them with strong affinity toward various nonpolar organic pollutants such as pyrene, phenanthrene, naphthalene, benzene, and so on [14]. The available π atomic electrons on individual graphene sheet of CNTs further help them to make a strong chemical complex with organic molecules having C=C bonds or benzene rings (such as polycyclic aromatic hydrocarbons (PAHs) and polar aromatic compounds) either through π–π stacking or π–π electron–donor–acceptor (EDA) interaction [15, 16]. Tournus and Charlier (while conducting ab initio studies on the interaction of benzene with CNTs) revealed that in the cases where π–π stacking interaction predominates, the adsorption behavior is greatly influenced by the chiral angle of the graphene sheets [17].

Organic compounds having carboxyl (–COOH), hydroxyl (–OH), or amino (–NH_2) functional groups also interact with the graphitic CNT surface mainly through hydrogen bonding where the latter acts as an electron donor [18]. Electrostatic attraction, generally offered by modified CNTs, facilitates the adsorption of positively charged organic chemicals such as some antibiotics (at suitable pH) and heavy metal ions. Covalent interactions, on the other hand, are predominant majorly with hydrophilic pollutants at the surface of essentially modified and multifunctional CNTs. Ji et al. [11] investigated the adsorption of tetracycline to CNTs, graphite, and AC and found that the adsorption affinity of tetracycline decreased in the order of graphite/SWNT > MWNT ≫ AC upon normalization for adsorbent surface area. The weaker adsorption of tetracycline to AC was greatly due to the truncated accessibility of available adsorption sites, whereas its remarkably strong adsorption to CNTs could be attributed to the strong adsorptive interactions (van der Waals forces, π–π EDA interactions, cation–π bonding) with the graphene surface of CNTs [11].

3.3.4 Tunable surface chemistry

The surface of the CNTs can be desirably modified so as to ensure better, selective, and productive interactions with different pollutants. In general, CNTs are hydrophobic in nature, and surface functionalization is performed chiefly to make them hydrophilic and water soluble. Functional groups such as – carboxyl (–COOH), hydroxyl (–OH), or carbonyl (–C=O) could be intentionally introduced onto CNT surfaces by acid oxidation or air oxidation (Figure 3.3). The functionalized CNTs so obtained aids in the adsorption of relatively low molecular weight and polar contaminants such as phenol [16] and 1,2-dichlorobenzene [14, 19] . In addition, the hydrophilic CNTs also enhance the ion-exchange rates and hence the adsorption capacity of CNTs in aqueous media [20]. Other methods of CNTs activation and modification involves combining with other metal ions [21] or metal oxides [22], and coupling with organic compounds [23]. In some cases, surface functionalization of CNTs could also be performed to make them selective for a particular type of pollutant. For instance, nitrogen- and boron-doped CNT surfaces were found to be better reactivated with Pt. It has been demonstrated that the adsorption rate of Pt was enhanced for both nitrogen- and boron-doped CNTs because of activating nitrogen neighboring carbon atoms and strong hybridization between the platinum d orbital and boron p orbital, respectively [20]. Besides, surface functionalization also provides a sustainable means to avoid aggregation of CNTs by generating repulsion between the individual tubes. Zhang et al. [24] highlighted that the aggregation of CNTs was unfavorable for the

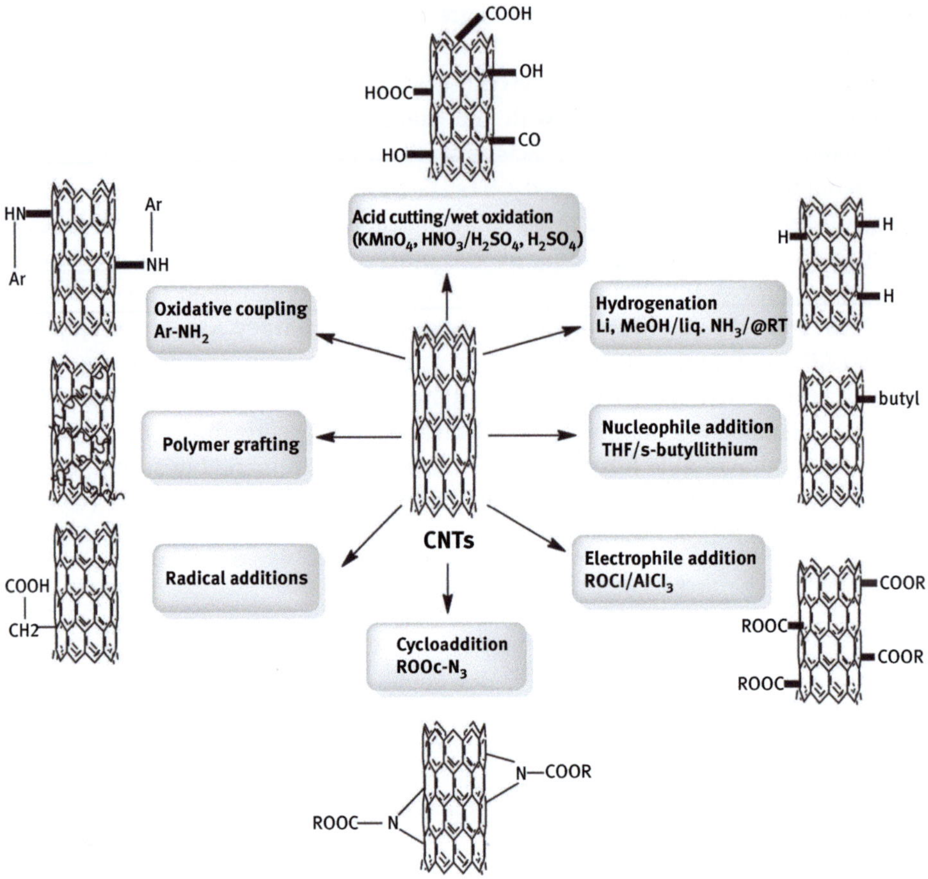

Figure 3.3: Various strategic routes for synthesizing functionalized CNTs.

adsorption of several synthetic organic compounds (SOCs such as pharmaceuticals, cosmetics) on CNTs as the surface area was investigated to be more important than the pore volume in adsorption of SOCs.

3.3.5 Broad-spectrum activity

The broad-spectrum activity of CNTs is a direct consequence of their tunable surface chemistry, diverse CNT–pollutant interactions, and unconventional adsorption sites such as intestinal spaces and peripheral grooves. Typically, the amounts of surface acidity (carboxylic, lactonic, and phenolic groups) favor the adsorption of polar compounds [25], whereas un-functionalized CNTs surface is proved to have higher adsorption capacity toward nonpolar compounds such as PAHs [26, 27]. In addition, the adsorption sites like peripheral grooves facilitate the adsorption of some deleterious pollutants such as pharmaceutical and microbes that are generally not adsorbed by the traditional adsorbents. Intestinal cavities also provide small-sized pores for the adsorption of some very small organic pollutants. As a consequence, CNTs display positive adsorption activity toward different classes of water pollutants.

A wealth of experimental data reveals that apart from the conventional organic, inorganic, and biological pollutants [8–14], CNTs have also shown great interest and potential in gas adsorption. Studies described CNT adsorption of gases such as ammonia [28], nitrogen and methane [29, 30], hydrogen [31–34], ozone [35], and carbon monoxide and carbon dioxide [36].

3.3.6 Flexible working conditions

Experimental investigations reveal that CNTs exhibit a wide working range in terms of temperature, pH, and ionic strength with optimum performance in the pH range of 7–10 [25]. Zhang et al. [24] observed that the solution's pH and ionic strength exhibited only slight or insignificant impacts on the adsorption of three representative SOCs (phenanthrene, biphenyl, and 2-phenylphenol), which vary drastically in planarity, polarity, and hydrogen/electron–donor/acceptor ability.

3.3.7 Possibility of regeneration and reuse

CNTs exhibit the extra edge advantage of recycling without significantly affecting their adsorption efficiency that is generally not shown by its other counterparts. Studies revealed that adsorption of metal ions on CNTs can be easily reversed by reducing the pH of the solution. The metal recovery rate is usually above 90% and often close to 100% at $pH < 2$ [37]. Moreover, the adsorption capacity remains relatively stable after regeneration. Lu et al. [38] reported that Zn^{2+} adsorption capacity of SWNT and MWNT decreased less than 25% after 10 regeneration and reuse cycles. In another study, it was demonstrated that the adsorption and desorption of Ni^{2+} in CNTs slightly decreased, but those of granular activated carbon (GAC) sharply decreased after a number of cycles. This phenomenon could be explained by the fact that the porous structure of GAC makes desorption of Ni^{2+} more difficult as the ions have to move from the inner surface to the external surface of the pore [39].

All these aforementioned factors make CNTs the most ideal candidate for the absorption processes till date. Owing to these prodigious advantages, they have been exploited severely for the removal of various classical and some really noxious pollutants in the recent times (Table 3.1). Additionally, the tunable surface chemistry and controllable pore size of CNTs make them a good support for composite adsorbents where they act as scaffold for macromolecules or metal oxides with intrinsic adsorption ability [14]. Some of the examples include CNT decoration with iron oxide for europium adsorption [40], chitosan for methyl orange adsorption [41], and ceria nanoparticles for chromium adsorption [42]. Moreover, the unique electrical properties of CNTs could be utilized for enhanced adsorption with electrochemical assist [43]. However at the same time, experimental studies have time and again emphasized that though the adsorption capacity of CNTs in static mode is incomparable, it does suffer from some efficiency drop in dynamic mode and is regulated by a number of parameters such as the nature of adsorbate, pH, temperature, presence of other pollutants, concentration of pollutants, contact time, size of particles, and so on [44]. Some of the parameters, namely, hydrophilicity of the CNTs' surface, might act a double-edged sword by simultaneously increasing and decreasing the adsorption capacity [8].

3.4 CNTs as nanofilter membranes

Membrane technologies, including microfiltration, ultrafiltration, nanofiltration, and reverse osmosis, have been established as a common means of purifying water long ago. Factors like absence of chemical additives, thermal inputs, and spent media regeneration make them more popular over other water treatment technologies [78]. Membrane technologies act by providing a physical and/or energy barrier to various pollutants and contaminants allowing the easy and uninterrupted flux of water molecules through them. The performance of these technologies is largely governed by the material constituting the membrane. A variety of materials ranging from polymeric (most common) to inorganic and even metallic have been employed for fabricating membranes for water purification processes. Most of these materials, however, suffer from the classical disadvantages of membrane filtration such as fouling, high energy consumption, optimum balance between retention–permeation capacity, and self-cleaning and reuse [4].

Well-aligned CNTs, in this context, serve as robust pores in membranes for water purification systems [79]. The exceptionally high aspect ratios of CNTs at such small dimensions and their molecularly slick, chemically inert hydrophobic graphitic walls act as collaterals for transport applications. Literature reports that liquid flow through membranes composed of an array of aligned CNTs is four to five times faster than that predicted from conventional fluid flow theory [80, 81]. Molecular dynamic simulations establish that such a dramatic increase in the flow is attributed to the hydrophobic nature of CNTs pores, which creates weak interactions with the water molecules, thereby enabling a fast and nearly frictionless flow of water [83]. Consequently, these membranes work with low energy consumption.

In addition, appropriate pore diameters of CNTs can constitute energy barriers at the channel entries, rejecting salt ions and permitting water through the nanotube hollows [82]. Thus, CNT membranes can be used as a “gate keeper” for size-controlled separation of multiple pollutants. It is also possible to modify CNT pores to selectively sense and reject ions [83]. Functionalities such as positive ($-NH^{3+}$), negative ($-COO^-$, sulfonic acids), and hydrophilic (aromatic) groups can be implanted at the tips and/or cores of CNTs by different wet oxidizing agent treatments that make them selective for particular pollutant retention and increase water influx through the nanotube holes [84, 85]. Besides, CNTs also offer antifouling, antimicrobial, self-cleaning, and reusable functions. Figure 3.4 highlights the superiority of CNTs membranes over other conventional membrane technologies [83].

Typically, there are two types of CNT membranes depending upon their fabrication process: (i) vertically aligned (VA) and (ii) mixed matrix (MM) CNT membranes. VACNTs membranes can be synthesized by aligning perpendicular CNTs with supportive filler contents (epoxy, silicon nitride, etc.) between the tubes. On the other hand, an MMCNT (also called nanocomposite-CNT hybrid membranes) membrane consists of several layers of polymers or other composite materials doped/treated with randomly aligned CNTs [86, 87]. In the case of VACNT membranes, the tubes are directly exposed to the liquid to filter and filtration occurs mainly via steric hindrance and follows the pore flow model (also referred as *sieving*). In contrast, CNTs in MM membranes (embedded in a polymer matrix) either act as internal “fast lanes” for water flow once the latter has entered the membrane or act as an adsorbent material for the contaminant when it forms a thin active film over the polymeric matrix (adsorptive and sieving mechanism) [86, 88, 89]. Both types of membranes provide powerful and reliable tools for the rejection of various and/or selective unwanted pollutants and spontaneous flux of water molecules together with decent antimicrobial activity. However, the tedious fabrication route of VACNTs makes them less popular in comparison to MMCNTs.

Apart from these membranes, buckypaper that is a self-supported mat of entangled CNTs (forming a flexible structure that is chemically and physically stable) is also known [90, 91]. It has attracted interest for use in filtration techniques due to its flexibility and simple fabrication process, relative to aligned CNT membranes [92, 93]. Early investigations into the permeability studies report that PVDF-grafted buckypapers were highly effective in removing bacteria and viruses from water supplies [94, 95]. Evidence also show that buckypapers could be used for desalination [93] or gas separation [96]. Recent research suggests that it may be possible to control the porosity of buckypapers by changing the average length of the MWNTs used in their preparation [97].

In membrane filtration technology, CNTs together with imparting its conventional benefits (of high selectivity, antimicrobial activity, and thermal and mechanical stability) are also recognized to prevent fouling of the bulk matrix membrane system. Generally membrane fouling takes place through two mechanisms, in which

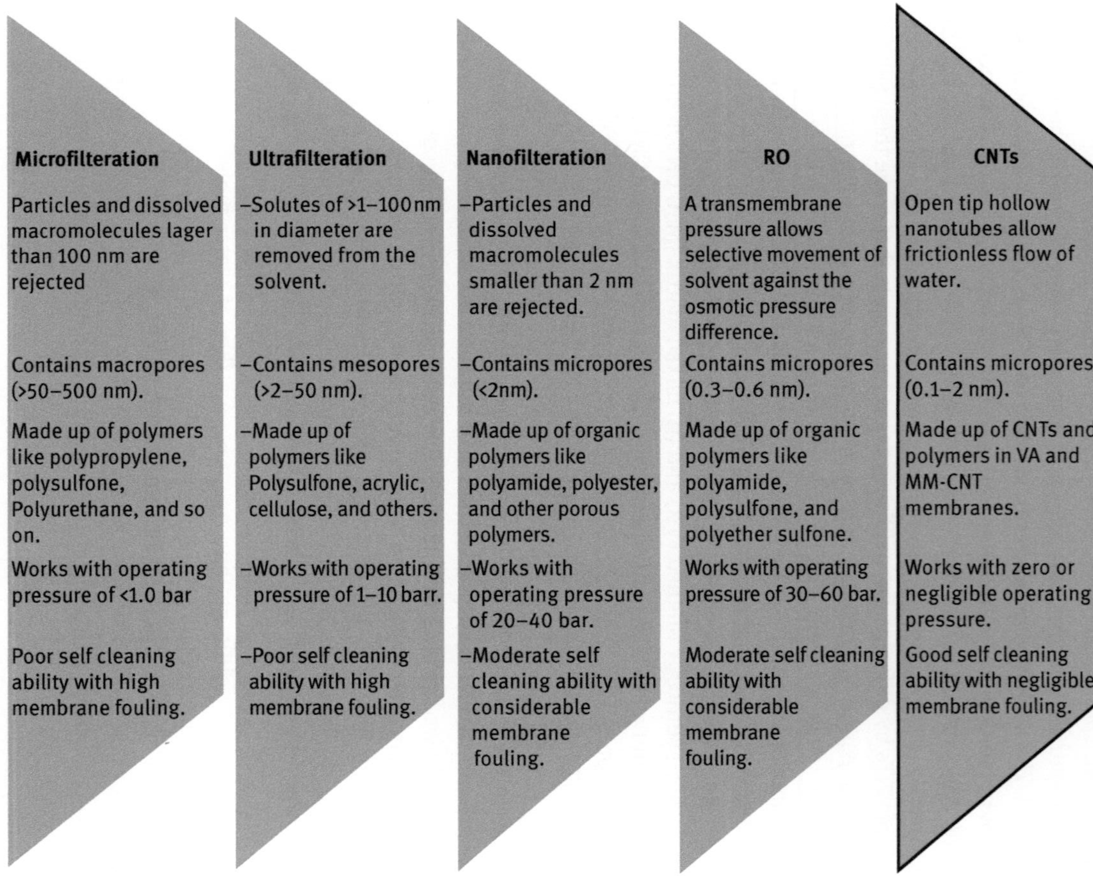

Figure 3.4: Comparative features of various membrane filtration technologies.

the first one occurs in membrane pores, while the second accumulates on the membrane surface due to various impurities, that is, suspended inorganic or organic materials, and biological and dissolved organic materials in the water influent [98]. Recent studies established that hydrophilic and smooth membrane surfaces enhance the membrane antifouling property due to the hydration via hydrogen bonding [99]. CNTs, although hydrophobic in nature, can be changed to hydrophilic moieties via acid treatment [14]. It has been demonstrated that functionalized CNT-blended polysulfone membrane [100] and polyethersulfone membrane [101] are more hydrophilic and have an enhanced fouling resistance due to the hydrophilic carboxylic groups of functionalized CNTs. Other functional groups, such as hydrophilic isophthaloyl chloride groups [102] and amphiphilic-polymer groups, can also be introduced onto CNT surface, which offers high protein-resistant ability [103]. Biofouling, the growth of a biofilm on membrane surfaces, can also be alleviated by the incorporation of CNTs where the latter act primarily as contact biocide [94, 104]. The biocidal activity of the MMCNTs membrane has been vastly shown to be boosted by grafting embedded CNTs with antibacterial silver ions [105, 106].

Literature cites arrays of studies that demonstrate the utilization of CNT membranes (VA, MM, and buckypaper) in various filtration processes ranging from simple desalination and removal of organic/inorganic (natural/dumped) pollutants from brackish water to effective separation of microorganisms (Table 3.2). CNTs have been efficiently doped with polymers like polysulfones, aromatic polyamides, chitosan, polyacrylonitrile, polyethersulfone, and cellulose acetate for achieving improved desalination and other filtration applications [100, 102, 107, 108].

Table 3.2: Potential of CNTs and its composites in removal of different water pollutants via membrane filtration.

Membrane type	Water pollutant	Comments	Ref.
VACNTs	Bovine serum albumin (BSA), colloidal silica, dextran, NaCl, and Na_2SO_4	Pure water permeability for the VACNT membrane higher 1000 ± 100 than UF membrane 400 L/m^2 h bar. Functionalization of VACNT membrane with methacrylic acid (MA) was attempted to not only retard membrane fouling, but also improve solute rejection than UF membrane.	[109]
DWCNT membrane	KCl and K_3FeCN_6	Rejection ability (%) for different ions was Cl^- (45), K^+ (37), $(FeCN_6)^{3-}$ (≈91), and 79% (3 K^+). Rejection mechanism was dominated by electrostatic interactions between fixed membrane charges and mobile ions, whereas steric and hydrodynamic effects appeared to be less important.	[110]

Table 3.2 (continued)

Membrane type	Water pollutant	Comments	Ref.
Multiwalled CNT immobilized membrane	NaCl	The membrane was highly stable over a long operational time. Carboxylated CNTs were incorporated into CNIM to enhance pure water flux. More than 99% of the salt reduction was achieved.	[111]
Polyvinylchloride (PVC)–MWCNTs–Co–Cu	Na^{2+}	Additive (0.5 to 8 wt.%) increased ionic permeability and flux. 0.5 wt.% additive made the ion transportation difficult and so declined the ionic permeability and flux.	[112]
MWNT–TiO_2 composites	Carbamazepine, ibuprofen, and acetaminophen	The peak initial removal percentages of the pharmaceuticals by the MWNT–TiO_2 membranes were 80%, 45%, and 24% for carbamazepine, ibuprofen, and acetaminophen, respectively. Peak removal efficiencies after regeneration were 55%, 32%, and 19% for carbamazepine, ibuprofen, and acetaminophen, respectively, indicating some loss in sorptive capacity upon regeneration.	[113]
SWCNTs coated polyvinylidine fluoride (PVDF) membrane	Viral and bacterial pathogens	Nearly 80% of the *E. coli* cells were inactivated after 20 min contact time (an eightfold increase over the uncoated PVDF membrane). Full virus removal (5–7 log removal) was observed with a 6 μm skin layer, whereas 3.2-log removal was seen with a thin 2 μm layer.	[114]
MWCNTs-cellulose acetate (CA) composite	NaCl	Permeation rates were found to improve by 54% with a minimal decrease in salt retention (–6%) for the MM membranes with the lowest CNT content. Further addition of CNTs caused a reduction in permeation rates, which is attributed to the decreased porosity and surface area.	[115]
Purified and oxidized buckypaper	Humic acid (HA)	The buckypaper exhibited excellent removal of HA (> 93%) and a long lifetime for filtration.	[116]

Table 3.2 (continued)

Membrane type	Water pollutant	Comments	Ref.
SWCNTs	Bacteriophage MS2	Solution's pH had a significant impact on viral removal as the interactions between viral particles and SWNTs changed from attractive below the virus isoelectric point (about pH 3.9) to repulsive at higher pH. The removal of viruses increased at higher ionic strengths (NaCl) due to suppression of repulsive electrostatic interactions between viruses and SWNTs.	[117]
Polyvinyl-*N*-carbazole (PVK)-SWCNT-coated nitrocellulose membrane	Gram-positive (*Bacillus subtilis*), Gram-negative (*E. coli*) bacteria and the model virus MS2.	Membranes coated with the nanocomposite exhibited significant antimicrobial activity toward Gram-positive and Gram-negative bacteria (~80–90%) and presented a virus removal efficiency of ~2.5 logs. PVK-SWNT (97:3 wt% ratio PVK:SWNT) coated filters will produce more suitable coated membranes for drinking water than pure SWNTs-coated membranes (100%), since the reduced load of SWNT in the nanocomposite would reduce the use of costly and toxic SWNT nanomaterial on the membranes.	[118]

3.5 CNTs as hybrid catalysts

Like adsorption and transport processes, CNTs have also shown profound ability in hybrid catalysis where they act as an active support for the conventional catalysts. The large SSA, sp^2 hybrid carbons, topological defects, high electrical conductivity, modifiable surface chemistry, and good thermal stability with sufficient resistance toward acidic and basic media are some of the merits of CNTs that contribute toward their superior cocatalytic efficiency [119]. Moreover, the uniform porous structure of CNTs further reduces the mass-transfer limitations of reactants from solution to active sites on the catalyst, thereby aids in higher catalytic output [14, 16]. As a consequence, CNTs have shown huge promise to actively participate in the photocatalytic, wet air oxidation catalysis and biocatalysis processes for wastewater purification in the past few decades.

3.5.1 CNTs in photocatalysis

Photocatalysis can be described as a photoreaction that is accelerated by light/ photon in the presence of single or multiple catalyst(s). The classical examples of photocatalysts include TiO_2, CdS, Fe_2O_3, ZnO, and so on [8]. They have been extensively used in the catalytic degradation of various organic pollutants in wastewater such as dyes, pesticides, pharmaceuticals, and so on for a very long time. However, their efficacy is strongly hampered by their low quantum efficiency, which in turn is due to the rapid recombination of photogenerated electrons and holes, that is, most charges quickly recombine without participating in the photocatalytic reactions [14].

CNTs are known to successfully enhance the quantum efficiency of these conventional catalysts by capturing electrons (excited to the conduction band) due to their low fermi level and high charge conducting ability that ultimately renders the recombination of electrons (excited to the conduction band) with holes (created in the valence band) [8]. The captured electrons (by CNTs) can then be transferred to another electron acceptor, such as molecular oxygen, forming reactive oxygen species (O_2^-, H_2O_2., and OH.) that degrade and further mineralize organic pollutants. Hence in a CNT–photocatalyst composite, CNTs can be called as a *photogenerated electron acceptor* that enhances interfacial electron transfer process, whereas conventional photocatalysts behave as good electron donors under irradiation [120–122]. Moreover, the holes created in the valence band could also oxidize the absorbed water, thereby forming the hydroxyl radical ($^{\bullet}OH$). The latter directly oxidize the adsorbed water pollutants on the CNT surfaces (Figure 3.5). Different functionalities on CNTs provide immense possibilities for doping them with different photocatalyitc agents [123] and hence accelerate versatile applications (Table 3.3).

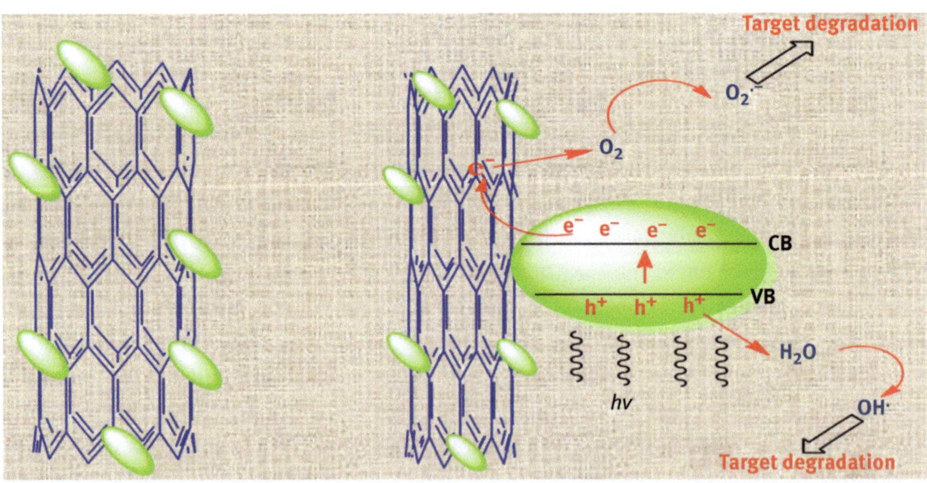

Figure 3.5: Mechanism of photocatalytic degradation of water pollutant by CNT nanocomposite.

Table 3.3: Potential of CNTs in degrading water pollutants via hybrid photocatalysis.

Catalyst nanocomposite	Water pollutant	Comments	Ref.
MWCNT/TiO_2	Methyl orange (MO)	MWCNT/TiO_2 mass ratio (1/2) showed maximum MO removal at 450 °C of sintering and pH 2. Increased reusability.	[124]
MWCNT/TiO_2	Bacillus cereus (B. cereus)	MWCNTs–TiO_2 showed more than three times larger specific surface area than commercial TiO_2. 90% inactivation of B. cereus was achieved. Solar UV lamps in the presence of MWCNTs–TiO_2 successfully inactivated the *B. cereus* while solar UV lamps only or solar UV lamps with Degussa P25 showed no significant inactivating behaviors.	[125]
Polyvinylalcohol (PVA)/ TiO_2/graphene–MWCNT	Methylene blue (MB)	PVA/TiO_2/graphene–MWCNT showed better photocatalytic activity in MB decomposition compared to PVA/TiO_2/graphene composite.	[53]
MWCNT/CdS	Azo dye	CNTs hampered the photocorrosion of CdS and hence enhanced the degradation rate.	[126]
MWCNT/TiO_2	Atrazine	Microwave was used to enhance the photocatalytic activity; CNT had a beneficial effect on absorbing microwave energy.	[127]
MWCNT/titanium silicate	4-Nitrophenol, Rhodamine B	Ball milling removes the physical contact between CNT and titanium silicate greatly reduced the photocatalytic activity, indicating the significance of interfacial charge transfer.	[128]
MWCNT/TiO_2	2,6-Dinitro-*p*-cresol	Enhanced photocatalytic efficiency of the composite in comparison to TiO_2 was observed. No obvious decline in quantum efficiency was observed after five repeated cycles.	[129]
CNT/mesoporous TiO_2	Acetone	Inhibition of charge recombination; more hydroxyl groups on the catalyst and more hydroxyl radicals generated.	[130]

Studies indicate that aqueous pollutants including dyes [53, 124], benzene derivatives [131], and carbamazepine [132] have been efficiently photodegraded by CNT–TiO_2 composite. Gray et al. [133] in their study for the photocatalytic efficiency of TiO_2/CNT nanocomposite revealed that the composite exhibited enhanced photocatalytic oxidation activity to phenol due to reduced charge recombination (as evidenced by the diminished photoluminescence intensity). In addition, SWCNT enhanced the photocatalytic activity of TiO_2 better than MWCNT owing to more individual contact between the SWCNT and the TiO_2 nanoparticle surface.

Besides, CNTs also possess ability to individually affect the photocatalytic efficiency of different photocatalysts. For instance, CNTs can tune the band gap of TiO_2 and hence can widen their absorption range from only ultraviolet to UV plus visible [134, 135]. The photocorrosion effect of CdS can also be minimized with the incorporation of CNTs as the latter could effectively adsorb the reducing agents in the solution (which can react with holes) and can stabilize CdS [126]. However, since CNTs can absorb the incident light, excess of CNTs may also induce adverse impact on the activity of the composite photocatalyst. Therefore, it is critical to control the dosage of CNTs to photocatalysts [130].

3.5.2 CNTs in catalytic wet air oxidation (CWAO)

CWAO can be described as a process where catalysts like Pt, Pd, and Ru utilize oxygen in the air to oxidize dissolved and suspended organic matter in water. CNT-supported Pt, Pd, and Ru catalysts have found intense applications in wet air oxidation of organic pollutants like phenol and aniline where the mesoporous nature of CNTs allow better diffusion of pollutants to the surface of catalyst [136–138]. Garcia et al. [139] reported the catalytic degradation of Azo dyes using CNT-supported Pt as the hybrid catalyst. The study emphasized that MWCNTs–Pt catalyst can significantly improve the total organic carbon and color removal efficiencies of textile effluents.

Besides, CNTs alone can also act as catalyst for wet air oxidation of organic pollutants. Yang et al. [140] discussed the application of functionalized MWCNTs for CWAO of phenol in a batch reactor. The study revealed that the acid-treated CNTs adsorb the liquid oxygen that was dissociated onto the graphitic sheets to form dissociated oxygen atom. The latter along with the carboxyl moiety on the CNT surface generates HO_2. radical through hydrogen bonding, which subsequently decomposed phenol to CO_2, H_2O, and low organic compounds in CWAO process [141] (Figure 3.6).

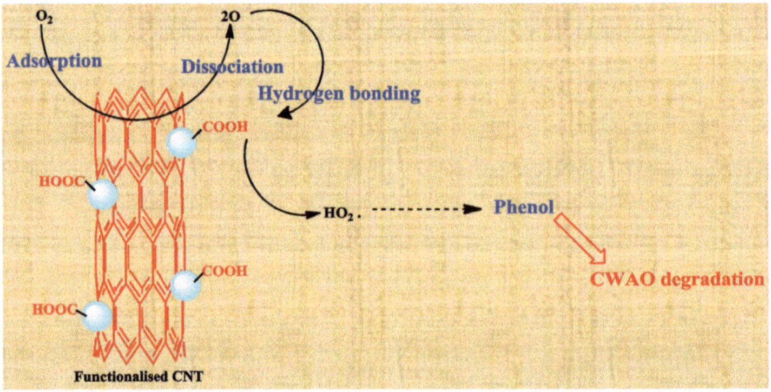

Figure 3.6: Mechanism of wet air oxidative degradation of phenol by MWCNT.

3.5.3 CNTs in biocatalysis

Enzymatic degradation of noxious compounds has drawn considerable attention in the last few years owing to the extra edge selectivity and performance provided by the enzymes over their chemical counterparts. The activity of naive enzymes however suffers largely from the issues like regeneration, narrow working range, and stability and, thus, augments the need of suitable solid supports for immobilization time and again. Recognizing the vital features provided by CNTs (such as large SSA and flexible working conditions), they have been recently utilized for the immobilization of enzymes so as to form entities called *nanobiohybrids*. Enzyme immobilization on the CNT surface could be achieved through three major routes, namely, binding to a support (physical adsorption and covalent bonding), cross-linking (carrier free), and encapsulation or entrapment [142]. In physical adsorption, spontaneous adsorption of enzymes onto CNTs surface occurs partially due to hydrophobic and electrostatic interactions [143]. Covalent bonding, on the other hand, is facilitated by CNT oxidation, followed by activation with carbodiimide chemistry [144, 145]. Cross-linking polymers such as chitosan, poly(diallyldimethylammonium chloride), and so on are used to immobilize enzymes on CNTs through cross-linking method [146]. Finally, a layer-by-layer approach is adopted for immobilizing enzymes through enzymes encapsulation [147].

Typically, nanobiohybrid catalysis is applied for water purification processes where other chemical transformations of water pollutants are not feasible. Other merits of this process include high catalytic activity of enzyme, prolonged mechanical stability, broad working conditions, stable and effective working in harsh reaction conditions, and most importantly reusability of the hybrid system, thereby making the process economically viable [146–149]. In general, the biohybrid can play three major functions with high selectivity and sensitivity in water purification processes: (i) binding and preconcentrating the pollutants; (ii) removal and/or degradation of pollutants; and (iii) sensing and monitoring the pollutants [8]. Table 3.4 provides an overview of the studies undertaken to explore the potential of CNTs-based nanobiohydrids [8].

Table 3.4: Potential of CNTs-based nanobiohydrids in water pollutant degradation.

Nanobiohybrid	Water pollutant	Comments	Ref.
Oxidized (O-MWCNTs)-laccase	ABTS,[a] bisphenol, and catechol	Loading capacity was highest for O-MWNTs and lowest for C_{60}. No obvious structural changes in the enzyme were observed after immobilization.	[150]
MWCNTs–tyrosinase	Phenol derivatives	The nanobiohybrid selectively oxidized phenol and its derivatives with extended stability of the catalyst at high temperatures.	[146]

Table 3.4 (continued)

Nanobiohybrid	Water pollutant	Comments	Ref.
SWCNTs–HRP[b] SWCNT–DM[c] Con A[d]- HRP	Dibenzothiophene (DBT)	DBT removal rate was 66.46 and 94.65% for SWCNT–HRP and SWNT–DM-Con A-HRP catalysts, respectively.	[151]
SWCNTs-1,2 HQD[e]	Catechol	Immobilization enhanced the working pH and temperature conditions of the enzyme and hence the degradation rate.	[152]
MWCNTs–SBP[f]	p-Cresol	The biohybrid maintained the structural integrity of the enzyme with remarkable enhancement in its activity and stability. Around 50% of the enzyme activity was retained after multiple uses.	[153]
SWCNTs–crude enzyme MWCNTs–crude enzyme	Aniline	SWCNTs–enzyme hybrid led to efficient removal of aniline in comparison to than MWCNTs–enzyme indicating effective reaction between aniline and enzymes on the surface of SWCNTs.	[154]

a 2, 2′-Azino-bis-(3-ethylbenzthiazoline-6-sulfonic acid) diammonium salt.
b Horseradish peroxidase.
c *n*-Dodecyl β-D-maltoside.
d Concanavalin A.
e Hydroxyquinol 1,2 dioxygenase.
f Soybean peroxidase.

3.6 CNTs in microbial fuel cells (MFC)

Microbial fuel cell (MFC) can be defined as a bioelectrochemical cell or system where current is produced by using microorganisms, mimicking natural bacterial interactions [8] (Figure 3.7). It is an environmental-friendly method of wastewater treatment with self-sustained generation of electric current through microorganisms. The system consists of two electrodes – anode and cathode in different compartments separated by a membrane. Anodic compartment, filled with wastewater, deals with the oxidation of organic matter (as the latter act as nutrient for the microorganisms), thereby producing electrons, protons, and CO_2. The produced electrons are then transferred to the anodic surface and finally to the cathode (terminal electron acceptor) by exoelectrogens (microorganisms that transfer electrons to an electrode). Exoelectrogens such as cytochromes exhibit redox potential at their outer membrane that mediates the transfer of electrons across the electrochemical cell. In the cathodic compartment, electrons are accepted usually by oxygen, the most sustainable terminal electron acceptor, thereby triggering oxygen reduction reactions [8, 14].

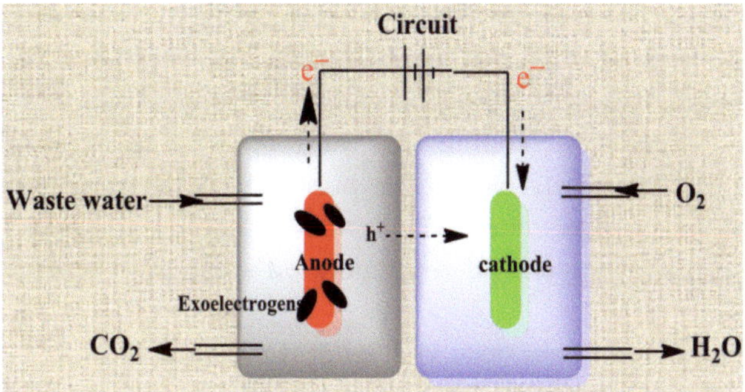

Figure 3.7: A schematic representation of a microbial fuel cell (MFC).

Since microbes play the most vital role in generation of electrons, an anode material that favors enhanced microbial colonization together with high conductivity is highly desirable. To date, various carbon-based materials such as carbon cloth, carbon paper, and carbon foam have been used as the anode material because of their ability to increase surface area (necessary for more bacterial attachment and colonization), current conductivity, stability, and catalytic activity [155]. However, the much superior characteristics of CNTs over the traditional ones make them the most promising anode material for MFC of the present times.

However despite the euphoria, CNTs also suffer from a major disadvantage of their cytotoxicity, which leads to proliferation inhibition and cell death and hence drastically hampers the bacterial growth [14]. To overcome this drawback, several polymers such as polyaniline [156], polypyrrole [157], and so on have been used for increasing CNT surface area and decreasing CNT antibacterial effects. In a recent study, Xie et al. [158] synthesized a biocompatible and highly conductive anode by using CNTs–textile composites in MFC. The anode, besides facilitating enhanced electrode–bacteria interactions, also offered lower charge-transfer resistance (10-fold) compared to traditional carbon cloth anode and, hence, higher current conductivity. Therefore, CNTs composite anode would have high electron density, low electrode decomposition, biocompatibility, high porous surface area, good mechanical contact, high chemical stability, and facilitator for transferring electrons in MFC [8].

3.7 CNTs in oil–water filtration

Besides the conventional organic and inorganic water pollutants, oily discharge into the water resources exists as the yet another major source of water pollution. Gallons of oily wastewater are generated every day from various industrial processes like metallurgical, transportation, food processing, and petrochemical as well as petroleum refineries [159]. The situation is further worsened by ill-fated oil spills into the water bodies that ultimately pose an alarming threat to the environment. Surveys have estimated that the United States generates over 2.4 billion gallons per day of oily waste water, which is equivalent to 7 barrels of contaminated water for each barrel of oil produced onshore. This amounts for nearly half of the public drinking water produced in the country [160, 161]. As a consequence, effective and scalable technologies for the removal of oil from water resources and reclaiming oil spills are earnestly required.

Current practices often use the gun-barrel (aka. knock-put wash) tank or its variations as the first step of oil–water separation under gravity. The water effluent of the gravity separation often contains oil droplets less than 10 μm in diameter as an oil–water emulsion [162]. Further treatment of the-oil polluted water involves the disruption of these oil/water emulsion in order to completely remove the oily discharge. Various technologies have been used to treat the oil/water emulsion, such as ultrasonic separation, coagulation, air flotation, heating, ozonation, flocculation, and membrane filtration [163–167]. By far, membrane filtration has shown superior efficacy over the other treatment methods in terms of its low operational pressure, chemical and thermal stability, and easy recyclability [168, 169]. Basically, it involves the water/oil emulsion filtration across a superhydrophbic and superoleophilic porous membrane such that it allows the easy permeation of oil while repelling the water molecules, respectively [170].

Both ceramic and metallic membranes (including Al_2O_3, Cu, TiO_2, ZrO_2, stainless steel (SS), etc.) have received considerable attention (over polymeric and cellulose-based membranes) in their applications for the oily wastewater treatment due to their porous structure along with high chemical, thermal, and mechanical stability [171]. However to suffice the aim meticulously, they require intensive modification (such as coating and grafting with hydrophobic molecules) to improve their hydrophobic and oleophilic properties. For example, coating of nanostructured copper mesh with long-chain fatty acids [172], and modification with poly(*N*-isopropylacrylamide) [173], has been normally performed. Coating the surface of SS mesh with a rough surface of poly(tetrafluoroethylene) has also been reported [174]. Organic and polymeric modifiers, however, are unable to withstand higher temperatures and conduct heat. They are even not very resistant to aggressive liquids such as selected organic solvents, salty, acidic, or basic aqueous solutions, common to many industrial emulsions, and hence are not commonly utilized for surface modifications [170].

CNTs, however, have been recently recognized as the favorable surface coating material over porous matrix membrane to heighten the performance of oil filtration membranes. This is because CNTs, besides exhibiting the unique characteristics of high tensile strength, chemical inertness, thermal stability, large surface area, and strong mechanical strength, are the naturally existing superhydrophobic and superoleophilic materials. Lee et al. [170] reported MW-VACNTs-coated stainless steel mesh as the efficient oil filtration membrane. Water advancing contact angles of 145–150° were determined for these SS-CNT meshes in air and oil (gasoline, isooctane). They found that the as-fabricated SS-CNT mesh has the ability to dewater water–oil emulsion and proposed that CNTs having higher affinity to oil than to water due to the hydrophobic interaction between oil molecules and the graphene sheets of CNTs allowed the efficient filtration of oil.

3.8 Graphene in wastewater treatment

In contrast to one dimensional (1D) CNTs, graphene is one-atom thick two-dimensional (2D) sheet of sp^2-bonded carbon atoms densely packed in the form of hexagonal honeycomb lattice with a carbon–carbon bond length of 0.142 nm [175]. It is the thinnest, strongest, and lightest material known till date. Since its discovery in 2004 by Novoselov, Geim, and coworkers, it has evoked tremendous interests in the scientific community around the world due to its unprecedented electrical, thermal, optical, and mechanical properties with low synthetic cost. It exhibits the most inimitable physico-chemical characteristics including high SSA of 2,600 $m^2\ g^{-1}$, excellent thermal conductivities of 5,000 $W\ m^{-1}\ K^{-1}$, high-speed electron mobility of 2,00,000 $cm^2\ V^{-1}\ s^{-1}$ at room temperature, high stiffness and strength with Young's modulus of around 1000 GPa and break strength of 130 GPa, and extraordinary catalytic activity [176–180]. These outstanding properties of graphene make them potential candidates for the applications in numerous areas including electronics, biomedical, sensing and monitoring, and environmental protection [181–184] However, their strong tendency to form irreversible agglomerates through hydrophobic interactions, namely, π–π stacking and van der Waals interactions limits their applicability. Consequently, special emphasis on the invention of methods required for surface functionalization (oxidation-reduction, thiolation, amidation, etc.) and modification (grafting, integration, intercalation, layer by layer deposition, etc.) of graphene sheets [185–187] has been laid to heighten their performance.

In the context of wastewater purification processes, graphene (and its derivatives) behave in the manner very similar to CNTs where it deals with the water contaminants via adsorption, rejection-permeation, and degradation mechanisms. Though the similarities between CNTs and graphene in terms of properties and mode of action are huge, a significant difference lies in the cost-effective fabrication of the former (even at larger scale), thereby making it more commercially viable. A wealth of investigations [188–192] has been cited in the literature that discusses the role of graphene in various wastewater treatment processes.

3.8.1 Graphene as adsorbents

As discussed in the case of CNTs, the excellent adsorption capacity of graphene sheets is largely due to the availability of enormous surface area and adsorption sites on the surface of graphene sheets. Since graphene-based nanomaterials are 2D structures, they expose much larger surface for adsorption in comparison to SWCNTs [190]. Further, the exposed surface can be easily tuned through wide range of surface modification and functionalization in order to induce both selectivity and versatility in the adsorption process. Consequently, arrays of water pollutants of distinct origin, namely, industrial, agricultural, and domestic can be efficiently adsorbed by graphene and its derivatives with fast adsorption kinetics.

Typically, pristine graphene sheets are hydrophobic in nature and hence could adsorb only nonpolar water pollutants through hydrophobic interactions. Functionalization of pristine graphene sheets is generally performed to make them more reactive, selective, biocompatible, soluble, and polar. Graphene oxide (GO) and reduced graphene oxide (RGO also called *graphene nanosheets* (GNs)) are the most explored examples of functionalized graphene nanomaterials. GO is characterized by the presence of plenty of oxygen-containing functional groups (such as –O–, –OH, and –COOH) and is synthesized by chemical oxidation and exfoliation of pristine graphite using either the Brodie, Staudenmaier, or Hummers method, or some variations of these methods [189]. GNs (or rGOs), on the other hand, are synthesized by the reduction of GOs with partial restoration of the native sp^2-hybridized network [193]. These nanomaterials are known to rapidly interact with contaminants (heavy metal ions, organic dyes, etc.) via diverse type of interactions, namely, π–π complexation, π–π stacking, hydrogen bonding, and electrostatic and covalent interactions [192].

Literature highlights remarkable ability of GOs in removal of inorganic ions (chiefly cations) due to their surface reactivity and negative charge distribution on their surface (Table 3.5). Recently, Zhao et al. conducted a sequence of experiments to unveil the adsorption capacity of few-layered GO nanosheets (GOS) for the removal of Cd^{2+} [194], Co^{2+} [194], Pb^{2+} [195], and U(VI) [196] ions from aqueous solutions. They found that the abundant oxygen-containing functional groups on GO surfaces played an important role in metal sorption allowing coordinate interactions with the metal ions. It was also revealed that solution parameters like ionic strength, pH, temperature, and presence of additives (humic acid) had significant effect on the adsorption capacity of GOS.

Exploiting the biosorbent ability of chitosan for metal ions removal, few synergistic nanocomposites of chitosan with GO have been explored [197–199]. All the studies highlighted that enhanced adsorption of the metal ions (Au^{3+}, Pb^{2+}, Pd^{2+}, Cu^{2+}) in the presence of GO together with rational regeneration and reuse of the composite was enabled. Similarly, nanocomposites of GO with magnetic nanoparticles have also been evaluated for their adsorption and regeneration efficiencies

Table 3.5: Potential of graphene-based nanomaterials in the adsorption of different water pollutants.

Adsorbent	Adsorbate	Comments	Ref.
		Removal of inorganic pollutants	
Graphene oxide (GO)	Cu^{2+}	Maximum adsorption capacity of 46.6 mg/g, higher than that of carbon nanotubes (28.5 mg/g) and active carbon (4–5 mg/g) for Cu^{2+} by GO was observed.	[213]
GO-SH	Hg^{2+}	GO-SH adsorbed sixfold higher concentration of Hg^{2+} ions than activated carbon.	[214]
GOs–EDTA silane	Pb^{2+}	A maximum adsorption capacity of 479 mg/g at pH 6.8 within 20 min of incubation was observed.	[215]
GO–gelatin/chitosan monoliths (CCGO)	Cu^{2+} and Pb^{2+}	Extremely high adsorbing capacity along with considerable recyclability of the composite (5 w% GO) was observed.	[198]
Chitosan/GO composites (CSGO)	Au^{3+} and Pd^{2+}	Maximum adsorption capacity of 1077 mg/g and 217 mg/g for Au^{3+} and Pd^{2+}, respectively, was observed at the initial loading capacity of 5 w% GO in the composite.	[199]
GO–TiO_2 hybrid	Zn^{2+}, Cd^{2+}, and Pb^{2+}	Adsorption capacities of 88.9 mg/g for Zn^{2+}, 72.8 mg/g for Cd^{2+}, and 65.6 mg/g for Pb^{2+}, respectively, at pH 5.6 was observed.	[216]
Fe_3O_4/GO composite	Co^{2+}	The composite showed higher adsorption for Co^{2+} ions than Fe_3O_4 and can be separated and recovered magnetically after adsorption. The adsorption capacity varied from 13.0 mg/g to 22.7 mg/g as temperature was increased from 30 °C–70 °C.	[201]
GNs	Pb^{2+}	GNs presented strong adsorption for Pb^{2+} ions due to their Lewis basicity.	[217]
GNs^{PF6} and GNs^{C8P}	Pb^{2+} or Cd^{2+}	The adsorption capacities of GNs^{PF6} were much higher than that GNs^{C8P} and carbon nanotubes. GNs^{PF6} showed highest adsorption capacity of 406.6 mg/g and 73.42 mg/g for Pb^{2+} and Cd^{2+}, respectively, in contrast to showed maximum adsorption capacity of 74.2 mg/g and 30.0 mg/g for Pb^{2+} and Cd^{2+}, respectively.	[218]

Table 3.5 (continued)

Adsorbent	Adsorbate	Comments	Ref.
SiO_2/graphene composite	Pb^{2+}	Maximum adsorption capacity for Pb^{2+} ion was 113.6 mg/g, which was much higher than that of bare SiO_2 nanoparticles. Electrostatic and hydrophobic interactions were deduced as the main driving force for the adsorption.	[219]
Polypyrrole (PPy)–rGO	Hg^{2+}	Maximum adsorption capacity of Hg^{2+} was 980 mg/g.	[220]
Magnetite–graphene hybrids	As^{3+} and As^{5+}	Approximately complete removal of arsenic ions (over 99.9%) within 1 ppb detection levels was observed.	[202]
Ethylenediamine (ED)–rGO	Cr^{6+}	π-Electrons of the carbocyclic six membered ring of ED–RGO aided in the effectual removal of chromium ions from aqueous solution through an indirect reduction process.	[204]
Fe_0/Fe_2O_3/Si–S–O–GNs	Cr^{6+}	Extremely fast removal (within 5 min of incubation) of the Cr(VI) ions with a high removal efficiency was observed.	[205]
Pristine graphene	F^-	An unexpected adsorption capacity of 35.6 mg/g at pH = 7.0 and 25 °C was observed.	[221]
Graphene–polypyrrole (Ppy) nanocomposite	ClO_4^-	The nanocomposite showed significantly improved adsorption for ClO_4^- compared with Ppy film alone.	[206]
		Removal of organic pollutants	
GO	Methylene blue (MB) and malachite green (MG),	The adsorption capacities of MB and MG on the GO were 351 and 248 mg/g, respectively, much higher than those on graphite and activated carbon. Electrostatic attractions between GO negative charge and cationic dyes were the main driving force.	[222]
3D GO sponge	Methylene blue (MB) and methyl violet (MV)	Maximum adsorption capacities of 397 mg/g and 467 mg/g for MB and MV, respectively, were observed. The activation energies data indicated strong π–π stacking as the most feasible mechanism for adsorption.	[207]

Table 3.5 (continued)

Adsorbent	Adsorbate	Comments	Ref.
GNs	Cationic red X-GR	The maximum monolayer adsorption capacity of 317 mg/g for red-X GR at 15 °C was obtained.	[223]
GO and in situ reduced rGO	Acridine orange (AO)	Under identical conditions, GO showed a maximum adsorption capacity of 1.4 g/g, and rGO provided a maximum adsorption capacity of 3.3 g/g, due to the conversion of carbonyl groups on GO into hydroxyl groups, thereby allowing electrostatic binding in the latter.	[224]
GOs and GNs	MB, MV, rhodamine B (RB), and an anionic dye, orange G (OG)	Where GOs were effective for adsorption of cationic dyes, GNs were found to show enhanced removal efficiency for anionic dyes. Electrostatic interaction took the hold for constructive adsorption of cationic dyes, whereas van der Waals forces accounted for the adsorption of OG on the surface of GNs.	[208]
MWCNT–graphene hybrid aerogel	Rhodamine B (RB), methylene blue (MB), fuchsine, and acid fuchsine	Adsorption capacities of graphene/c-MWCNT composite were 150.2, 191.0, 180.8, and 35.8 mg/g for RB, MB, fuchsine, and acid fuchsine, respectively.	[225]
(3D) chitosan–graphene mesostructures	Reactive black 5 (RB5)	A removal efficiency of 97.5% was achieved readily within few minutes of treatment.	[226]
GN-Ti hybrids	Methylene blue (MB)	Maximum adsorption at the rate of 83.3 mg/g over the hybrid in comparison to pure GNs (48.7 mg/g) and tubular titanates (36.5 mg/g) was achieved.	[227]
GO–Fe_3O_4	MB and neutral red (NR)	The adsorption capacities for MB and NR were 167.2 and 171.3 mg/g, respectively.	[228]
$Fe_3O_4^-$ graphene	Fuchsine	Maximum adsorption efficiency of 89.4 mg fuchsine/g of composite, with noteworthy regeneration of the adsorbent was achieved.	[229]
$CoFe_2O_4$–GNs	Methyl orange (MO)	An adsorption capacity as high as 71.5 mg/g was achieved.	[230]

Table 3.5 (continued)

Adsorbent	Adsorbate	Comments	Ref.
rGO supported ferrite composites	RB and MB	Removal efficiency of 92% and 100% for RB and MB, respectively, within 2 min of incubation at an initial composite conc. of 0.6 g/L was achieved.	[231]
GNs	Acrylonitrile (AN), p-toluenesulfonic acid (p-TA), 1-naphthalenesulfonic acid (1-NA) and methyl blue (MB).	The maximum adsorption capacities of GNs for p- TA, 1-NA, and MB were 1430, 1460, and 1520 mg/g (at 30 °C), respectively. Pollutant with a greater number of benzene rings showed higher adsorption, chiefly due to the enhanced hydrophobic interactions between the two.	[232]
GNs	Phenol	The maximum adsorption capacity of GNs with adsorption rate of 28.26 mg/g at incubation conditions pH 6.3 and temperature 12 °C were achieved.	[233]
Sand–graphene composite	Rhodamine 6G (R6G) and chlorpyrifos (CP)	Adsorption capacities of 55 mg/g for R6G and 48 mg/g for CP were achieved.	[234]
Sulfonated GNs	Naphthalene and 1-naphthol	The adsorption capability of sulfonated GNs was almost thrice (2.3–2.4 mmol/g) as that obtained from activated carbon and MWCNTs. Strong π–π interaction between benzene rings of the pollutants and π net of the graphene sheets contributed to the high adsorption.	[235]
GO/polypyrrole (PPy) composites	Phenol and aniline	The sorption of phenol and aniline on GO/PPy composites was mainly attributed to ion exchange, π–π electron donor–acceptor (EDA) interaction, hydrophobic interaction, and Lewis acid–base interaction.	[236]

PF6: potassium hexafluorophosphate;
C8P: 1-octyl-3-methyl-imidazolium hexafluorophosphate;
LDH: layered double hydroxides.

[189]. In general, incorporation of magnetic particles in GOs combats their separation problem and decreases the possibility of serious agglomeration and restacking of the graphene sheets, thereby providing higher available surface area for adsorption [200]. Liu et al. [201] reported a magnetite Fe_3O_4/GO composite (M/GO) for the removal of Co^{2+} from aqueous solutions. It was found that M/GO showed higher

adsorption efficiency than that on Fe_3O_4 and can be separated and recovered by magnetic separation after adsorption.

Unlike GOs, GNs display relatively lower adsorption affinity due to scarcity of functional groups (thereby reduced charge density) at their surface postreduction. However, the large surface area of graphene sheets enables the adsorption of various heavy metal ions (Pb^{2+}, Cd^{2+}, Hg^{2+}, Ni^{2+}) and anions (F^-, PO_4^{3-}) mainly via π–π interactions and van der Waals forces [189]. Magnetic nanocomposites of GNs have also been found to be effective against efficient removal of some highly deleterious ions like arsenic [202, 203] and chromium [204, 205] where the driving force for the pollutant removal was deduced to be predominantly hydrophobic π–π interactions. Zhang et al. [206] reported a 3D nanostructured graphene–polypyrrole (Ppy) nanocomposite for perchlorate (ClO_4^-) purification. The graphene–Ppy nanocomposite exhibited a significantly improved uptake capacity for ClO_4^- ions compared with Ppy film alone and can be used for ClO_4^- removal through an electrically switched ion exchange.

Apart from inorganic ions, graphene nanomaterials have also been known to competently adsorb organic pollutants such as dyes, pesticides, phenolic compounds, and so on. Most of the organic dyes dissolve in water and are usually present either in cationic or anionic form in effluent water. Investigations revealed that GO exhibits stronger affinity toward cationic dyes (due to the negative charge on the GO surface with predominant electrostatic interactions), whereas GNs show comparable affinity toward both cationic and anionic dyes with predominantly π–π stacking interactions acting between the two [189]. Plethora of experimental studies has time and again authenticated the splendid adsorption ability of graphene nanomaterials toward the removal of organic dyes (Table 3.5). Liu et al. [207] recently made a progressive step by synthesizing a 3D GO sponge for the removal of cationic dyes, methylene blue (MB), and methyl violet (MV). The 3D GO sponge demonstrated adsorption capacities of 397 and 467 mg/g for MB and MV, respectively. In another study, Ramesha et al. [208] compared GO and GNs for the adsorption of cationic dyes, MB, MV, rhodamine B (RB), and an anionic dye, orange G (OG) from aqueous solutions. It was found that GO showed stronger affinity for MB, MV, and RB (due to large negative charge density on their surface), whereas GNs were found to be a very good adsorbent for anionic dyes. Electrostatic interactions were the predominant driving force for the removal of MB and MV, while van der Waals forces accounted for the removal of OG by GNs.

Besides organic dyes, the adsorption of SOCs was also investigated. Gao et al. [209] studied the adsorption of antibiotics tetracycline, oxytetracycline, and doxycycline on GO and found the maximum adsorption capacities of 313, 212, and 398 mg/g, respectively. Similarly, Pei et al. [210] examined the adsorption extent of 1,2,4-trichlorobenzene (TCB), 2,4,6-trichlorophenol (TCP), 2-naphthol, and naphthalene on GO and GNs. They deduced that while TCB, TCP, and 2-naphthol adsorption on GNs was mainly due to π–π interaction, the adsorption of TCP and 2-naphthol on GO was attributed to the formation of H-bonds between hydroxyl groups (of TCP and 2-naphthol) and O-containing functional groups on GO.

Graphene nanomaterials have also been demonstrated as efficient adsorbents for the removal of pesticides. Maliyekkal et al. [211] reported the exceptional adsorption of chlorpyrifos (CP), endosulfan (ES), and malathion (ML) on GO and GNs from aqueous solution. They revealed that GNs exhibited better adsorption than GO with adsorption capacities of 1,200, 1,100, and 800 mg/g for CP, ES, and ML, respectively. Density functional analysis emphasized that adsorption onto GNs was mediated chiefly through water as the direct interactions between GNs and the pesticides were rather weak or unlikely. Recently, graphene-based materials have also been applied for the removal of organic and inorganic pollutants existing simultaneously in wastewater [212], thereby highlighting their potential in a more realistic environment. Table 3.5 provides a summarized view of the potential of graphene based nanomaterials in the adsorption of different water pollutants.

3.8.2 Graphene in membrane filters

Graphene has been envisaged as an excellent starting material for developing separation membrane owing to its unique characteristics of superior mechanical strength [178], high tensile strength, atomic thickness [237], relative chemical inertness, and hydrophobicity [238–240]. However, negligible permeability of the water molecules associated with the pristine graphene sheets (due to its densely packed honeycomb lattice structure) restricts their utility as nanofilters for water treatment in wastewater purification processes [190]. Generation of pores with well-defined sizes into the graphene basal planes has been seen as the most prompted solution of this limitation [241]. The nanopores, introduced into the sheets, act as nanocapillaries/channels for almost frictionless flow of water through them. Controlled formation of subnanometer sized pore into the graphene sheets was first achieved by ion bombardment of graphene layer with gallium ions followed by oxidative etching of the sheet so as to promote growth of generated defects into controlled pores [190]. Various other techniques for pores generation have also been developed including electron beam [242], block copolymer [243] and nanosphere lithographies [244], barrier-guided chemical vapor deposition [245], photocatalytic oxidation [246] catalytic hydrogenation [247], and chemical etching [248]. Graphene sheets with nanopores so formed are generally referred as *nanoporous graphene* (NPGs) [241].

NPGs have been projected as the most promising size-selective free standing membrane material till now [239]. Molecular dynamic simulations as well as experimental studies [249–251] have established that NPGs can exhibit high permeability and selectivity that is with few orders of magnitude higher than those of the existing state-of-the-art membranes [252, 253].The performance of NPGs-based membranes in efficient rejection (salt ions and other organic pollutants) and permeation (of water molecules) is largely governed by their pore structure (in terms of size, volume, and density), chemical functionalization, and the operating conditions [254]. Existence of several surface functionalization, modification, and intercalation strategies bestows plenty of possibilities for the fabrication of functionalized nanopores that act as ionic sieves for selective retention of ions and other organic pollutants. Therefore in a precisely arranged array of functionalized nanopores (in graphene sheet), while the pore size deals with the stearic rejection of the ions (and other organic matter), the functionality at the pores determines the ion selectivity on the basis of electrostatic and hydration effects [255].

The pioneer work of using NPGs for separation membranes was attempted by Aluru and coworkers in 2010 through computational studies [256]. They recognized that graphene monolayers can allow water molecules to flow through them while hindering the passage of other unwanted species provided pore sizes and functional groups available at the nanopores have been carefully tailored. They also revealed that pure water can continually flow across graphene pores with diameters smaller than 1 nm. In a similar kind of work, Grossman et al. [253] revealed that NPG membrane could achieve more than 99% salt rejection and can provide water transport of up to 66 L/cm^2·day·MPa. Later, different research groups have synthesized graphene sheet functionalized with carboxylate, hydroxyl, and amine-terminated polystyrene to facilitate spontaneous water transport and selective ion rejection [257, 258]. In a recent study, Striolo and coworkers [252] investigated the effect of different functional groups on the transport properties of NPGs. They observed that hydroxyl-functionalized NPGs demonstrate maximum water flux due to the generation of strong energy barrier for Cl^- ions at both low and moderate ionic strengths as well as by accomplishing water substitution within the hydration shell of salt ions leading to fast water permeation [190].

Apart from NPGs, graphene derivatives, namely, GO and rGO have also been visualized as potential membrane materials. The functional groups on the surface of GO layers (epoxy, hydroxyl, and carboxyl) prop open the layers and permeation occurs mainly through percolating regions of "empty space" (where the sheets are not oxidized) between the layers [259]. As a consequence, enhanced water transport together with rejection of undesired moieties through size exclusion or charge effect mechanisms [260] could be achieved. In a recent study, Joshi et al. [261] revealed that GO sheets can act as molecular sieves when immersed in water and reject all solutes with a hydrated radii larger than 4.5 A°. In another study, O'Hern et al. [262] showed that the selectivity of GO membranes can be tuned by introducing sub-nm-sized pores through oxidative etching method. They demonstrated that the pores were cation selective when oxidative etching time was short, indicating electrostatic repulsions as the major driving force. However, the pores prevented the transport of larger organic molecules at longer oxidative etching time pointing toward steric size exclusion as the predominant mechanism of rejection.

Despite several benefits of NPGs-based free standing membrane materials, they suffer from some irresistible drawbacks such as poor thermal and chemical stability along with low mechanical strength (due to generation of defects), which confine their practical applicability [190]. Consequently, combining the graphene nanomaterials with suitable supports (polymeric and ceramic) so as to fabricate nanocomposite membranes (with polymers and ceramic host matrix) with improved physiochemical properties becomes an underlying need to exploit the captivating features of graphene nanomaterials in wastewater treatment processes. Nanocomposite membranes usually show superior activity in comparison to both the conventional counterparts as well as against their single components (of polymers or the graphene nanomaterial) with the extent of improvement depending upon the degree of modification [190]. Akin et al. [263] synthesized a nanocomposite membrane of polysulfone, embedded with polyaniline conjugated rGO sheets, for the selective permeation of water with considerable salt ions rejection. It was found that incorporation of rGO into the membrane matrix led to expanded porosity and enhanced macrovoid substructure. Further, the hydrophilic nature of the polyaniline-rGo conjugate allowed higher water flux through the membrane, achieving a maximum of 82% NaCl rejection.

Effect of cross-linked GO sheets onto the separation characteristics of polymeric host matrix polydopamine-coated polysulfone was also investigated [185]. The nanocomposite was fabricated via layer-by-layer deposition of cross-linked GOS onto the polymeric support. It was found the 2D channels formed between the stacked GO layers (with desirable spacing) and the charges on individual GOS played crucial roles in water permeation with subsequent rejection of unwanted ions and organic species. Moreover, the hydrophilic and antibacterial nature of GO sheets [264, 265] served as barriers to obstruct the adsorption of hydrophobic foulants (and microorganisms) as well as to retard the active chlorine species from diffusing toward the core area of the selective layers, hence preventing the fouling (and biofouling) along with deterioration (by chlorine) of the host TFC membrane to a larger extent [266].

Modified GO nanocomposites have also been known for their noticeable potential in oil/water filtration processes. Hu and colleagues in their attempt to separate oil from water synthesized GO-coated Al_2O_3 ceramic microfiltration membranes. The water permeation fluxes of unmodified and GO modified membranes were deduced to be around 522 and 667 $l\ h^{-1}\ m^{-2}\ bar^{-1}$ after 150 min, respectively, indicating the exceptionally high tendency of GO in oil filtration [267]. Moreover, in a recent study, Xue et al. [268] discovered a new member of the graphene nanomaterials family called *graphyne*, a one-atom thick carbon allotrope of graphene, prepared by replacing certain carbon–carbon bonds in graphene with acetylenic linkages. The nanomaterial showed 100% salt rejection with water permeability that was two orders of magnitude higher than commercially available RO membranes. Due to such fascinating “trade-off” mechanism, it is expected to show huge promises in the separation processes in the coming years.

3.8.3 Graphene in hybrid photocatalysis

Graphene is a zero band gap semiconductor with high electron conductivity. Upon integration with other semiconductors, it is known to significantly enhance the photocatalytic activity of the parent semiconductor chiefly by alleviating the migration efficiency of the photoinduced electrons (due to its π–π conjugation net and high electrical conductivity) and retarding the recombination of exciton pair [192]. Various hybrid systems of conventional photocatalyst with graphene such as TiO_2/graphene [269–273], CdS/graphene [274, 275], ZnO–graphene [276–278], $BiVO_4$/graphene [279], XFe_2O_4/graphene (X = Zn, Mn, Co) [280–282], CuO/ graphene [283], Cu_2O/graphene [284], Mn_2O_3/graphene [285], Ag/AgY/GO [286], and so on have been investigated and they show considerable photocatalytic activity toward degradation of different types of water pollutants. In the hybrid systems, graphene not only acts as a photogenerated electron acceptor for the excited electrons but also provides sufficient space for the adsorption of organic pollutants, hence synergistically enhancing the overall quantum efficiency of the nanocomposite. Table 3.6 provides a brief outline of the significant studies carried out toward assessing the potential of graphene nanomaterials in hybrid photocatalysis.

Table 3.6: Potential of graphene-based nanomaterials in pollutant degradation by hybrid photocatalysis.

Hybrid catalyst system	Water pollutant	comments	Ref.
Graphene–TiO_2 (P25)	Methylene blue	Introduction of graphene enhanced the degradation rate of MB from only 25% (in 55 min) by bare P25 to 85% (in 55 min) by the dyad composite. Photocatalytic degradation was also observed under visible light with 65% dye degradation in 65 min.	[269]
Graphene–TiO_2	Rhodamine B	G-TiO_2 showed enhanced degradation rate of RB from only 30% (in 30 min) by bare TiO_2 to 85% (in 30 min) by the hybrid system.	[270]
Graphene–TiO_2	Malachite green	Degradation rate by the composite (0.0674 min^{-1}) was higher than that by the semiconductor alone (0.0218 min^{-1}).	[271]

Table 3.6 (continued)

Hybrid catalyst system	Water pollutant	comments	Ref.
Graphydyne–TiO_2 and G-TiO_2	Methylene blue	The degradation rate of MB by graphydyne-TiO_2, graphene-TiO_2, and bare TiO_2 was 0.0247 min^{-1}, 0.0195 min^{-1}, and 0.0152 min^{-1}, respectively	[272]
CeO_2^- TiO_2^- graphene (CT-GH), CeO_2–TiO_2–carbon nanotubes (CT–CNTs) and CeO_2–TiO_2–active carbon (CT–AC)	Reactive Red 195 (RR195) and 2,4-dichlorophenoxyacetic acid (2,4-D)	Under UV irradiation, maximum degradation rate of RR195 and 2, 4-D was provided by CT-GH, mainly due to the availability of large surface are in comparison to CNTs and lower electron–hole recombination rate relative to AC.	[273]
Graphene–BiOBr	Rhodamine B	Upon visible light irradiation, the composite showed enhanced degradation rate of RB from only 0.042 min^{-1} by bare TiO_2 to 0.132 min^{-1} by the hybrid system.	[287]
Graphene–ZnS	Methyl orange (MO)	Upon irradiation with mercury lamp, the composite showed enhanced degradation of MO (0.02558 min^{-1}) in comparison to bare ZnS (0.00609 min^{-1}).	[274]
Graphene–Bi_2S_3	Methyl orange (MO)	Upon irradiation with mercury lamp, the composite showed enhanced degradation of MO (0.01204 min^{-1}) in comparison to bare Bi_2S_3 (0.00734 min^{-1}).	[274]
Graphene–CdS	Methyl orange (MO)	Upon irradiation with mercury lamp, the composite showed enhanced degradation of MO (0.01362 min^{-1}) in comparison to bare CdS (0.00481 min^{-1}).	[274]

Table 3.6 (continued)

Hybrid catalyst system	Water pollutant	comments	Ref.
Graphene–macro-CdS	Rhodamine B	G/M-CdS photocatalysts show superior photoactivity in degradation of Rh B under visible light irradiation.	[275]

Besides, graphene in hybrid photocatalytic systems is also known to affect individual catalytic attributes. For instance, graphene could decrease the band gap of TiO_2 (P25) nanoparticles such that they behave as effective visible light catalyst in conjugation with graphene [277]. ZnS nanoparticles also extend their absorption edge toward the visible range in conjugation with graphene [274].

3.9 Future prospects and conclusion

In the last few decades, tremendous efforts have been devoted to tackle the issue of water scarcity by developing several strategies for purifying wastewater. Although progress in the field was noteworthy, yet anticipated performance could not be achieved. The advent of "nano" era in the water-reframing program opened new areas of research and experimentation and hailed the remarkable potential of carbon nanomaterials in water treatment technologies. The unique and astonishing physicochemical aspects of carbon nanomaterials bestow them with numerous prospects of practical applications (as adsorbents, filters, catalysts, etc.) for the treatment of wastewater in the coming years. They are perfectly suitable for POU purposes as they constitute some of the lightest, compact, and tremendously effective tools for fresh water recovery from polluted water. Carbon nanomaterials (and their composites) with desirable selectivity, stability, strength, and reactivity can be certainly fabricated that widens their applicability range and open new avenues of multifunctional materials. Moreover, CNTs and graphene-based nanomaterials can be easily grafted/integrated/intercalated with other classical supports to harness maximum of their captivating features. In certain cases, they even present a synergistic response by enhancing the intrinsic properties of hybrid nanomaterial (as in the case of hybrid catalysis). Plenty of surface functionalization and manipulation approaches are available that can tune the nanomaterial characteristics as desired. Filtration membranes based on carbon nanomaterials provide myriad of opportunities against the other state-of-the-art membranes and introduce an altogether different genre of filtration technology. In addition, the antibacterial property of carbon nanomaterials endows them with the capability of overcoming the most severe drawback of traditional materials, which is fouling and pore blocking of the active sites. Hence, carbon nanomaterials can be looked upon as highly promising materials for providing new dimensions in the wastewater purification technology.

However despite the euphoria, the journey of carbon nanomaterials from the level of fundamental research (in laboratories) to pilot scale and commercial market is not easy. At present, the utility of carbon nanomaterials (at practical level) in wastewater purification is in very nascent stages and a lot more studies need to be done to evaluate their potential under more realistic conditions. Factors such as material performance under diverse quality of influent water, effects of microorganisms other than bacteria and virus, long-term efficacy, assessment of compounds produced during treatment, and so on need to be addressed before their large-scale applications could be realized. Collective and concerted solutions for different technical hurdles, such as control of pore size in NPGs and CNTs, restoration of chirality in CNTs, effective loading (in case of nanocomposites) during fabrication process, and the associated expenses that can undermine the performance of these nanomaterials, are also required. Finally, since the future of carbon nanomaterials depends largely on their cost-effective and quality-controlled fabrication together with minimum of their potential threat toward environment and human health, intense focus on the regeneration and reuse of the nanomaterials (which reduces the overall cost) along with negligible leaching into the environment is required so as to relish their productive and progressive impact for a long time. Hence, it can be established that carbon nanomaterials undoubtedly excel over traditional materials in terms of efficiency and versatility, but significant efforts toward their synthesis, maintenance, and toxicity are required to practically integrate them with the traditional water treatment processes. Moreover, though relatively a large number of studies on the potential of CNTs in wastewater treatment have been performed, the aura of graphene nanomaterials is still concealed. Consequently, a parade of studies need to done to completely define the boundaries of carbon nanomaterials in water purification processes. To conclude, it can be very well said that these black nanomaterials exhibit remarkable aptitude to mark the beginning of a golden era in the wastewater purification arena.

References

[1] UN-Report, International Decade for Action "Water for Life" 2005–2015, Available at: http://www.un.org/waterforlifedecade/scarcity.shtml 2012 (Retrieved on: 05, May, 2015).

[2] World Health Organization (WHO), World Health Report: Reducing Risks, Promoting Healthy LifeFrance. Available at: http://www.who.int/whr/2002/en/whr02_en.pdf 2002 (Retrieved: 14th May 2015).

[3] Wastewater Management A UN-Water Analytical Brief. Available at http://www.unwater.org/fileadmin/user_upload/unwater_new/docs/UN-Water_Analytical_Brief_Wastewater_Management.pdf

[4] Xiaolei, Q., Pedro, J.J., and Alvarez, Q.L., Water Research, 2013, 47, 3931.
[5] Upadhyayula, V.K.K., Deng, S.G., Mitchell, M.C., and Smith, G.B. Science of the Total Environment, 2009, 408, 1.
[6] Khin, M.M., Nair, A.S., Babu, V.J., Murugan, R., and Ramakrishna, S. Energy and Environment Science, 2012, 5, 8075.
[7] Ong, Y.T., Ahmad, A.L., Zein, S.H.S., and Tan, S.H. Brazilian Journal of Chemical Engineering, 2010, 27(2), 227.
[8] Das, R., Bee, S., Hamida, A., Ali, Md. E., Ismail, A.F., Annuar, M.S.M., and Ramakrishna, S. Desalination, 2014, 354, 160.
[9] Yang, K., and Xing, B.S. Chemical Reviews, 2010, 110(10), 5989.
[10] Pan, B., and Xing, B.S. Environmental Science and Technology, 2008, 42(24), 9005.
[11] Ji, L.L., Chen, W., Duan, L., and Zhu, D.Q. Environmental Science and Technology, 2009, 43(7), 2322.
[12] Zhang, S., Shao, T., Kose, H.S., and Karanfil, T. Environmental Toxicology and Chemistry, 2012, 31(1), 79.
[13] Lu, C.S., Chung, Y.L., and Chang, K.F. Water Research, 2005, 39(6), 1183.
[14] Liu, X., Wang, M., Zhang, S., and Pan, B. Journal of Environmental Sciences, 2013, 25(7), 1263.
[15] Chen, W., Duan, L., and Zhu, D.Q. Environmental Science and Technology 2007, 41(24), 8295.
[16] Lin, D.H., and Xing, B.S. Environmental Science and Technology, 2008, 42(19), 7254.
[17] Tournus, F., and Charlier, J.C. Physical Review B, 2005, 71, 16.
[18] Yang, K., Wu, W.H., Jing, Q.F., and Zhu, L.Z. Environmental Science and Technology, 2008, 42(21), 7931.
[19] Peng, X.J., Li, Y.H., Luan, Z.K., Di, Z.C., Wang, H.Y., Tian, B.H., and Jia, Z.P. Chemical Physics Letters, 2003, 376(1–2), 154.
[20] Li, Y.H., Hung, T.H., and Chen, C.W. Carbon, 2009, 47, 850.
[21] Gupta, V.K., Agarwal, S., and Saleh, T.A. Journal of Hazardous Materials, 2011, 185, 17.
[22] Zhao, X., Jia, Q., Song, N., Zhou, W., and Li, Y. Journal of Chemical & Engineering Data, 2010, 55, 4428–4433.
[23] Kosa, S.A., Al-Zhrani, G., and Salam, M.A. Chemical Engineering Journal, 2012, 181–182, 159.
[24] Zhang, S.J., Shao, T., Bekaroglu, S.S.K., and Karanfil, T. Environmental Science and Technology, 2009, 43(15), 5719.
[25] Rao, G.P., Lu, C., and Su, F. Separation and Purification Technology, 2007, 58(1), 224.
[26] Wang, X., Jialong, L.U., and Xing, B. Environmental Science and Technology, 2008, 42(9), 3207.
[27] Yang, K., Zhu, L., and Xing, B. Environmental Science and Technology, 2006, 40(6), 1855.
[28] Bauschlicher, C.W., and Ricca, A. Physical Review B, 2004, 70, 115409.
[29] Bienfait, M., Zeppenfeld, P., Pavlovsky, N.D., Muris, M., Johnson, M.R., Wilson, T., DePies, M., and Vilches, O.E. Physical Review B, 2004, 70(3), 035410.
[30] Talapatra, S., and Migone, A.D. Physical Review B, 2002, 65(4), 045416.
[31] Dillon, A.C., Jones, K.M., Bekkedahl, T.A., Kiang, C.H., Bethune, D.S., and Heben, M.J. Nature, 1997, 386, 377.
[32] Chen, P., Wu, X., Lin, J., and Tan, K.L. Science, 1999, 285, 91.
[33] Liu, C., Fan, Y.Y., Liu, M., Cong, H.T., Cheng, H.M., and Dresselhaus, M.S. Science, 1999, 286, 1127.
[34] Lee, S.M., and Lee, Y.H., Applied Physics Letters 2000, 76, 2877.
[35] Yim, W.L., and Liu, Z.F. Chemical Physics Letters, 2004, 98, 297.
[36] Varghese, O.K., Kichambre, P.D., Gong, D., Ong, K.G., Dickey, E.C., and Grimes, C.A. Sensors Actuators B, 2001, 81(1), 32.

[37] Li, Y.H., Di, Z.C., Ding, J., Wu, D.H., Luan, Z.K., and Zhu, Y.Q. Water Research, 2005, 39(4), 605.
[38] Lu, C.S., Chiu, H., and Liu, C.T. Industrial & Engineering Chemistry Research, 2006, 45(8), 2850.
[39] Lu, C., Liu, C., and Rao, G.P. Journal of Hazardous Materials, 2008, 151(1), 239.
[40] Chen, C.L., Wang, X.K., and Nagatsu, M. Environmental Science and Technology, 2009, 43(7), 2367.
[41] Zhu, H.Y., Jiang, R., Xiao, L., and Zeng, G.M. Bioresource Technology, 2010, 101(14), 5063.
[42] Di, Z.C., Ding, J., Peng, X.J., Li, Y.H., Luan, Z.K., and Liang, J. Chemosphere, 2006, 62(5), 861.
[43] Li, X.N., Chen, S., Quan, X., and Zhang, Y.B. Environmental Science and Technology, 2011, 45(19), 8498.
[44] Sun, W., and Zhou, K. Chemical Engineering Journal, 2014, 258, 185.
[45] Bai, Y.C., Lin, D.H., Wu, F.C., Wang, Z.Y., and Xing, B.S. Chemosphere, 2010, 79(4), 362.
[46] Mohammed, M.I., Razak, A.A.A., and Al-Timimi, D.A.H. Advances in Materials Science and Engineering, 2014, 2014. http://dx.doi.org/10.1155/2014/201052
[47] Ji, L.L., Shao, Y., Xu, Z.Y., Zheng, S.R., and Zhu, D.Q. Environmental Science and Technology, 2010, 44(16), 6429.
[48] Sheng, G.D., Shao, D.D., Ren, X.M., Wang, X.Q., Li, J.X., Chen, Y.X., and Wang, X.K. Journal of Hazardous Materials, 2010, 178(1–3), 505.
[49] Wanga, S., Ng, C.W., Wang, W., Li, Q., and Hao, Z. Chemical Engineering Journal, 2012, 197, 34.
[50] Mohammadi, A., Kazemipour, M., Ranjbar, H., Walker, R.B., and Ansari, M. Fullerenes Nanotubes Carbon Nanostructures, 2015, 23, 165.
[51] Kumar, S., Bhanjana, G., Jangra, K., Dilbaghi, N., and Umar, A. Journal of Nanoscience and Nanotechnology, 2014, 14, 4331.
[52] Szlachta, M., and Wójtowicz, P. Water Science and Technology, 2013, 68(10), 2240.
[53] Jung, G., and Kim, H.I. Journal of Applied Polymer Science, 2014, 131, 40715.
[54] Mahmoodi, N.M., Bagherpour, F., and Nariyan, E. Desalination and Water Treatment, 2014, 1.
[55] Jadhav, A.H., Mai, X.T., Ofori, F.A., and Kim, H. Chemical Engineering Journal, 2014, 259, 348.
[56] Prola, L.D.T., Machado, F.M., Bergmann, C.P., de Souza, F.E., Gally, C.R., Lima, E.C., Adebayo, M.A., Dias, S.L.P., and Calvete, T. Journal of Environmental Management, 2013, 130, 166.
[57] Li, Y.H., Wang, S., Wei, J., Zhang, X., Xu, C., Luan, Z., Wu, D., and Wei, B. Chemical Physics Letters, 2002, 357, 263.
[58] Kandah, M.I., and Meunier, J.L. Journal of Hazardous Materials, 2007, 146, 283.
[59] Chen, C., Hu, J., Shao, D., Li, J., and Wang, X. Journal of Hazardous Materials, 2009, 164, 923.
[60] Li, Y.H., Wang, S.W., Luan, Z.L., Ding, J.D., Xu, C.X., and Wu, D. Carbon, 2003, 41, 1057.
[61] Tofighy, M.A., and Mohammadi, T. Journal of Hazardous Materials, 2011, 185, 140.
[62] Vukovic, G.D., Marinkovic, A.D., Colic, M., Ristic, M.D., Aleksic, R., Grujic, A.A.P., and Uskokovic, P.S. Chemical Engineering Journal, 2010, 57, 238.
[63] Wu, C.H. Journal of Colloid and Interface Science, 2007, 311(2), 338.
[64] Lu, C., Liu, C., and Su, F. Desalination, 2009, 249, 18.
[65] Atieh, M.A. Procedia Environmental Sciences, 2011, 4, 281.
[66] Li, Y.H., Wang, S.G., Cao, A.Y., Zhao, D., Zhang, X.F., Xu, C.L., Luan, Z.K., Ruan, D.B., Liang, J., Wu, D.H., and Wei, B.Q. Chemical Physics Letters, 2001, 350, 412.
[67] Peng, X.J., Luan, Z.K., Ding, J., Di, Z.H., Li, Y.H., and Tian, B.H. Materials Letters, 2005, 59(4), 399.
[68] Somera, B.F., Corazz, M.Z., Yabe, M.J.S., Segatelli, M.G., Galunin, E., and Tarley, C.R.T. Water, Air and Soil Pollution, 2012, 223, 6069.
[69] Deb, A.K.S., Ilaiyaraja, P., Ponraju, D., and Venkatraman, B. Journal of Radioanalytical and Nuclear Chemistry, 2012, 291, 877.
[70] Rehman, S., Ullah, N., Kamali, A.R., Ali, K., Yerlikaya, C., and Rehman, H. New Carbon Materials, 2012, 27(6), 409.
[71] Cui, Y., Liu, S., Hu, Z.J., Liu, X.H., and Gao, H.W. Microchimica Acta, 2011, 174, 107.

[72] Rananga, L., and Magadzu, T. Interaction of silver doped carbon nanotubes, Water Science and Technology: Water Supply, 2013, 14(3), 367.
[73] Liu, C., Xie, X., Zhao, W., Liu, N., Maraccini, P.A., Sassoubre, L.M., Boehm, A.B., and Cui, Y., Nano Letters 2013, 13, 4288.
[74] Sui, M., Zhang, L., Sheng, L., Huang, S., and She, L. Science of the Total Environment, 2013, 452, 148.
[75] Lilly, M., Dong, X., McCoy, E., and Yang, L. Environmental Science and Technology, 2012, 46, 13417.
[76] Sheng, L., Huang, S., Sui, M., Zhang, L., She, L., and Chen, Y. Frontiers of Environmental Science and Engineering, 2015, 9, 625.
[77] Zardini, H.Z., Amiri, A., Shanbedi, M., Maghrebi, M., and Baniadam, M. Colloids and Surfaces. B: Biointerfaces, 2012, 92, 196.
[78] Pendergast, M.M., and Hoek, E.M. Energy and Environment Science, 2011, 4, 6, 1946.
[79] Elimelech, M., and Phillip, W.A. Science, 2011, 333(6043), 712.
[80] Holt, J., Park, H., Wang, Y., Stadermann, M., Artyukhin, A., Grigoropoulos, C., Noy, A., and Bakajin, O. Science, 2006, 312, 1034.
[81] Noy, A., Park, H.G., Fornasiero, F., Holt, J.K., Grigoropoulos, C.P., and Bakajin, O. Nano Today, 2007, 2(6), 22.
[82] Corry, B. Journal of Physical Chemistry B, 2008, 112(5), 1427.
[83] Das, R., Ali, Md. E., Bee, S., Hamida, A., Ramakrishna, S., and Chowdhury, Z.Z. Desalination, 2014, 336, 97.
[84] Goh, P., Ismail, A., and Ng, B. Desalination, 2013, 308, 2.
[85] Kar, S., Bindal, R., and Tewari, P. Nano Today, 2012, 7, 385.
[86] Ahn, C.H., Baek, Y., Lee, C., Kim, S.O., Kim, S., Lee, S., Kim, S.H., Bae, S.S., Park, J., and Yoon, J. Journal of Industrial & Engineering Chemistry, 2012, 18, 1551–1559.
[87] Hinds, B.J., Chopra, N., Rantell, T., Andrews, R., Gavalas, V., and Bachas, L.G. Science, 2004, 303, 62.
[88] Mattian, D., Leese, H., and Lee, K.P., Journal of Membrane Science, 2015, 475, 266.
[89] Vecitis, C.D., Schnoor, M.H., Rahaman, S., Schiffman, J.D., and Elimelech, M. Environmental Science and Technology, 2011, 45(8), 3672.
[90] Baughman, R.H., Cui, C., Zakhidov, A.A., Iqbal, Z., Barisci, J.N., Spinks, G.M., et al. Science, 1999, 284(5418), 1340.
[91] Sˇpitalsky, Z., Aggelopoulos, C., Tsoukleri, G., Tsakiroglou, C., Parthenios, J., Georga, S., et al. Materials Science and Engineering B: Solid State Materials for Advanced Technology, 2009, 165(3), 135.
[92] Viswanathan, G., Kane, D.B., and Lipowicz, P.J. Advanced Materials, 2004, 16(22), 2045.
[93] Dume´e, L.F., Sears, K., Schu¨ tz, J., Finn, N., Huynh, C., Hawkins, S., et al. Journal of Membrane Science, 2010, 351(1–2), 36.
[94] Brady-Estevez, A.S., King, S., and Elimelech, M. Small, 2008, 4, 481.
[95] Brady-Estevez, A.S., Schnoor, M.H., Vecitis, C.D., Saleh, N.B., and Elimelech, M. Langmuir, 2010, 26, 14975.
[96] Smajda, R., Kukovecz, A., Konya, Z., and Kiricsi, I. Carbon, 2007, 45, 1176.
[97] Kukovecz, A., Smajda, R., Konya, Z., and Kiricsi, I. Carbon, 2007, 45, 1696.
[98] Amiri, M.C., and Samiei, M. Desalination, 2007, 207, 361.
[99] Tu, S.C., Ravindran, V., Den, W., and Pirbazari, M. AIChE Journal, 2001, 47, 1346.
[100] Choi, J.H., Jegal, J., and Kim, W.N. Journal of Membrane Science, 2006, 284(1–2), 406.
[101] Celik, E., Park, H., Choi, H., and Choi, H. Water Research, 2011, 45(1), 274.
[102] Qiu, S., Wu, L.G., Pan, X.J., Zhang, L., Chen, H.L., and Gao, C.J. Journal of Membrane Science, 2009, 342(1–2), 165–172.

[103] Liu, Y.L., Chang, Y., Chang, Y.H., and Shih, Y.J. ACS Applied Materials & Interfaces, 2010, 2(12), 3642.
[104] Tiraferri, A., Vecitis, C.D., and Elimelech, M. ACS Applied Materials & Interfaces, 2011, 3(8), 2869.
[105] Gunawan, P., Guan, C., Song, X.H., Zhang, Q.Y., Leong, S.S.J., Tang, C.Y., Chen, Y., Chan-Park, M.B., Chang, M.W., Wang, K.A., and Xu, R. ACS Nano, 2011, 5(12), 10033.
[106] Booshehri, A.Y., Wang, R., and Xu, R. Chemical Engineering Journal, 2013, 230, 251.
[107] Shawky, H.A., Chae, S.R., Lin, S., and Wiesner, M.R. Desalination, 2011, 272, 46.
[108] Tang, C., Zhang, Q., Wang, K., Fu, Q., and Zhang, C. Journal of Membrane Science, 2009, 337, 240.
[109] Park, S.M., Jung, J., Lee, S., Baek, Y., Yoon, J., Seo, D.K., and Kim, Y.H. Desalination, 2014, 343, 180.
[110] Fornasiero, F., Park, H.G., Holt, J.K., Stadermann, M., Grigoropoulos, C.P., Noy, A., and Bakajin, O. Proceedings of National Academy of Sciences, 2008, 105, 17250.
[111] Bhadra, M., Roy, S., and Mitra, S. Separation and Purification Technology, 2013, 120, 373.
[112] Zendehnam, A., Mokhtari, S., Hosseini, S., and Rabieyan, M. Desalination, 2014, 347, 86.
[113] Zaib, Q., Mansoor, B., and Ahmad, F. Environmental Sciences: Processes Impacts, 2013, 15, 1582.
[114] Ajmani, G.S., Goodwin, D., Marsh, K., Fairbrother, D.H., Schwab, K.J., Jacangelo, J.G., and Huang, H. Water Research, 2012, 46, 5645.
[115] Badawi, N.E., Ramadan, A.R., Esawi, A.M.K., and El-Morsi, M. Desalination, 2014, 344, 79.
[116] Yang, X., Lee, J., Yuan, L., Chae, S.-R., Peterson, V.K., Minett, A.I., Yin, Y., and Harris, A.T. Carbon, 2013, 59, 160.
[117] Brady-Este´vez, A.S., Nguyen, T.H., Gutierrez, L., and Elimelech, M. Water Research, 2010, 44, 3773.
[118] Ahmed, F., Santo, C.M., Mangadlao, J., Advincula, R., and Rodrigues, D.F. Water Research, 2013, 47, 3966.
[119] Eder, D. Chemical Reviews, 2010, 110, 1348.
[120] Saleh, T.A., and Gupta, V.K. Journal of Colloid and Interface Science, 2011, 362, 337.
[121] Riggs, J.E., Guo, Z., Carroll, D.L., and Sun, Y.P. Journal of American Chemical Society, 2000, 122, 5879.
[122] Subramanian, V., Wolf, E.E., and Kamat, P.V. Journal of American Chemical Society, 2004, 126, 4943.
[123] Saleh, T.A., The Role of Carbon Nanotubes in Enhancement of Photocatalysis: Syntheses and Applications of Carbon Nanotubes and Their Composites, Eds., S. Suzuki InTech publishers, Croatia, 2013, DOI: 10.5772/51050.
[124] Ma, G., Zhu, Y., Zhang, Z., and Li, L. Applied Surface Science, 2014, 313, 817.
[125] Lee, S.H., Pumprueg, S., Moudgil, B., and Sigmund, W. Colloids and Surfaces. B: Biointerfaces, 2005, 40, 93.
[126] Ma, L.L., Sun, H.Z., Zhang, Y.G., Lin, Y.L., Li, J.L., Yu, K.W.Y., Yu, Y., Tan, M., and Wang, J.B. Applied Surface Science, 2008, 19, 11.
[127] Chen, H.Z., Yang, S.G., Yu, K., Ju, Y.M., and Sun, C. Journal of Physical Chemistry A, 2011, 115(14), 3034.
[128] Krissanasaeranee, M., Wongkasemjit, S., Cheetham, A.K., and Eder, D. Chemical Physics Letters, 2010, 496(1–3), 133–138.
[129] Wang, H., Wang, H.L., and Jiang, W.F. Chemosphere, 2009, 75(8), 1105.
[130] Yu, Y., Yu, J.C., Yu, J.G., Kwok, Y.C., Che, Y.K., Zhao, J.C., Ding, L., Ge, W.K., and Wong, P.K. Applied Catalysis, A: General, 2005, 289(2), 186.
[131] Silva, C.G., and Faria, J.L. Applied Catalysis, B: Environmental, 2010, 101(1–2), 81.

[132] Martinez, C., Canle, M., Fernandez, M.I., Santaballa, J.A., and Faria, J. Applied Catalysis, B: Environmental, 2011, 102(3–4), 563.
[133] Yao, Y., Li, G., Ciston, S., Lueptow, R.M., and Gray, K.A. Environmental Science and Technology, 2008, 42(13), 4952.
[134] Lu, S.Y., Tang, C.W., Lin, Y.H., Kuo, H.F., Lai, Y.C., Tsai, M.Y., Ouyang, H., and Hsu, W.K. Applied Physics Letters, 2010, 96, 231915
[135] Cong, Y., Li, X.K., Qin, Y., Dong, Z.J., Yuan, G.M., Cui, Z.W., and Lai, X.J. Applied Catalysis, B: Environmental, 2011, 107(1–2), 128.
[136] Yang, S.X., Li, X., Zhu, W.P., Wang, J.B., and Descorme, C. Carbon, 2008, 46(3), 445.
[137] Taboada, C.D., Batista, J., Pintar, A., and Levec, J. Applied Catalysis B: Environmental, 2009, 89(3–4), 375.
[138] Garcia, J., Gomes, H.T., Serp, P., Kalck, P., Figueiredo, J.L., and Faria, J.L. Carbon, 2006, 44(12), 2384.
[139] Garcia, J., Gomes, H., Serp, P., Kalck, P., Figueiredo, J., and Faria, J. Catalysis Today, 2005, 102, 101.
[140] Yang, S., Zhu, W., Li, X., Wang, J., and Zhou, Y. Catalysis Communications, 2007, 8, 2059.
[141] Rivas, F., Kolaczkowski, S., Beltran, F., and McLurgh, D. Chemical Engineering Science, 1998, 53, 2575.
[142] Hwang, E.T., and Gu, M.B. Engineering in Life sciences, 2013, 13, 49.
[143] Nepal, D., and Geckeler, K.E. Small, 2006, 2, 406.
[144] Huang, W., Taylor, S., Fu, K., Lin, Y., Zhang, D., Hanks, T.W., Rao, A.M., and Sun, Y.P. Nano Letters, 2002, 2, 311.
[145] Jiang, K., Schadler, L.S., Siegel, R.W., Zhang, X., Zhang, H., and Terrones, M. Journal of Materials Chemistry, 2004, 14, 37.
[146] Subrizi, F., Crucianelli, M., Grossi, V., Passacantando, M., Pesci, L., and Saladino, R. ACS Catalysis, 2014, 4, 810.
[147] Feng, W., and Ji, P. Biotechnology Advances, 2011, 29, 889.
[148] Brena, B., González-Pombo, P., and Batista-Viera, F. Immobilization of enzymes: A literature survey, Immobilization of Enzymes and Cells Eds., J.M. Guisan, Humana Press, New York, USA, 2013, 15.
[149] Asuri, P., Bale, S.S., Karajanagi, S.S., and Kane, R.S. Current Opinion in Biotechnology, 2006, 17, 562.
[150] Pang, R., Li, M., and Zhang, C. Talanta, 2015, 131, 38.
[151] Li, Y., Huang, X., and Qu, Y. Journal of Chemical Technology and Biotechnology, 2013, 88, 2227.
[152] Suma, Y., Kim, D., Lee, J.W., Park, K.Y., and Kim, H.S. Proceedings of the International Conference on Chemical, Environmental Science and Engineering (ICEEBS'12), 2012.
[153] Asuri, P., Karajanagi, S.S., Sellitto, E., Kim, D.Y., Kane, R.S., and Dordick, J.S. Biotechnology and Bioengineering, 2006, 95, 804.
[154] Yan, H., Yang, X., Chen, J., Yin, C., Xiao, C., and Chen, H. Journal of Environmental Sciences, 2011, 23, 1165.
[155] Liu, X., Wang, M., Zhang, S., and Pan, B. Journal of Environmental Sciences, 2013, 25, 1263.
[156] Qiao, Y., Li, C.M., Bao, S.J., and Bao, Q.L. Journal of Power Sources, 2007, 170, 79.
[157] Zou, Y., Xiang, C., Yang, L., Sun, L.X., Xu, F., and Cao, Z. International Journal of Hydrogen Energy. 2008, 33, 4856.
[158] Xie, X., Hu, L., Pasta, M., Wells, G.F., Kong, D., Criddle, C.S., and Cui, Y. Nano Letters, 2010, 11, 291.
[159] Li, L., Ding, L., Tu, Z., Wan, Y., Clausse, D., and Lanoisellé, J. Journal of Membrane Science, 2009, 342, 70.

[160] Veil, J.A., Puder, M.G., Elcock, D., and Redweik, R.J. 2004. A White Paper Describing Produced Water from Production of Crude Oil, Natural Gas, and Coal Bed Methane. Argonne National Laboratory. Available at http://www.ipd.anl.gov/anlpubs/2004/02/49109.pdf (retrieved at 15th may 2015)
[161] Clark, C.E., and Veil, J.A. 2009. Produced Water Volumes and Management Practices in the United States. Argonne National Laboratory. Available at http://www.ipd.anl.gov/anlpubs/2009/07/64622.pdf (retrieved at 15th may 2015)
[162] Wanga, H., Lin, K.-Yi., Jing, B., Krylova, G., Sigmon, G.E., McGinn, P., Zhu, Y., and Naa, C. Water Research, 2013, 47, 4198.
[163] Stacka, L.J., Carney, P.A., Malone, H.B., and Wessels, T.K. Ultrasonics Sonochemistry, 2005, 12, 153.
[164] Bensadok, K., Belkacem, M., and Nezzal, G. Desalination, 2007, 206, 440.
[165] Binner, E.R., Robinson, J.P., Kingman, S.W., Lester, E.H., Azzopardi, B.J., Dimitrakis, G., and Briggs, J. Energy Fuels, 2013, 27, 3173.
[166] Kiss, Z.L., Kocsis, L., Szabó, G.K., Hodúr, C., and László, Z. Desalination and Water Treatment, 2014, 6, 1.
[167] Zhong, J., Sun, X., and Wang, C. Separation and Purification Technology, 2003, 32, 93.
[168] Darvishzadeh, T., Tarabara, V.V., and Priezjev, N.V. Journal of Membrane Science, 2013, 447, 442.
[169] Hemmati, M., Rekabdar, F., Gheshlaghi, A., Salahi, A., and Mohammadi, T. Desalination and Water Treatment, 2012, 39, 33.
[170] Lee, C.H., Johnson, N., Drelich, J., and Yap, Y.K. Carbon, 2011, 49, 669.
[171] Abadi, S.R.H., Sebzari, M.R., Hemati, M., Rekabdar, F., and Mohammadi, T. Desalination, 2011, 265, 222.
[172] Wang, S., Song, Y., and Jiang, L. Nanotechnology, 2007, 18(1), 015103.
[173] Song, W., Xia, F., Bai, Y., Liu, F., Sun, T., and Jiang, L. Langmuir, 2006, 23(1), 327
[174] Feng, L., Zhang, Z., Mai, Z., Ma, Y., Liu, B., Jiang, L., and Zhu, D. Angewandte Chemie International Edition, 2004, 43(15), 2012.
[175] Slonczewski, J.C., and Weiss, P.R. Physical Review, 1958, 109, 272.
[176] Yan, L., Zheng, Y.B., Zhao, F., Li, S., Gao, X., Xu, B. et al., Chemical Society Reviews, 2012, 41, 97.
[177] Lu, Y.H., Chen, W., Feng, Y.P., and He, P.M. Journal of Physical Chemistry B, 2009, 113.
[178] Lee, C.G., Wei, X.D., Kysar, J.W., and Hone, J. Science, 2008, 321, 385.
[179] Singh, A.K., and Yakobson, B.L. Nano Letters, 2009, 9, 1540.
[180] Stankovich, S., Dikin, D.A., Dommett, G.H.B., Kohlhaas, K.M., Zimney, E.J., Stach, E.A. et al., Nature, 2006, 442, 282.
[181] Bai, H., and Li, C., Shi, G. Advanced Materials, 2011, 23, 1089.
[182] Zhu, Y.W., Murali, S., Cai, W., Li, X., Suk, J. W., Potts, J. R., Ruoff, R. S. et al. Advanced Materials, 2010, 22, 3906.
[183] Huang, X., Yin, Z., Wu, S., Qi, X., He, Q., Zhang, Q., Yan, Q., Boey, F., and Zhang, H. Small, 2011, 7, 1876.
[184] Abergel, D.S.L., Apalkov, V., Berashevich, J., Ziegler, K., and Chakraborty, T. Advances in Physics, 2010, 59, 261.
[185] Dreyer, D.R., Park, S., Bielawski, C.W., and Ruoff, R.S. Chemical Society Reviews, 2010, 39, 228.
[186] Ramanathan, T., Abdala, A.A., Stankovich, S., Dikin, D.A., Herrera-Alonso, M., Piner, R.D., Adamson, D.H., Schniepp, H.C., Chen, X., Ruoff, R.S., Nguyen, S.T., Aksay, I.A., Prud'Homme, R.K., and Brinson, L.C. Nature Nanotechnology, 2008, 3, 327.
[187] Huang, X., Qi, X.Y., Boeya, F., and Zhang, H. Chemical Society Reviews, 2012, 4, 666.
[188] Mishra, A.K., and Ramaprabhu, S. Desalination, 2011, 282, 39.

[189] Wang, S., Sun, H., Ang, H.M., and Tadé, M.O. Chemical Engineering Journal, 2013, 226, 336.
[190] Goh, P.S., and Ismail, A.F. Desalination, 2015, 356, 115–128.
[191] Varma, S., Sarode, D., Wakale, S., Bhanvase, B.A., and Deosarkar, M.P. International Journal of Chemical and Physical Sciences, 2013, 2,132.
[192] Wang, H., Yuan, X., Wu, Y., Huang, H., Peng, X., Zeng, G., Zhong, H., Liang, J., and Ren, M. Advances in Colloid and Interface Science, 2013, 195–196, 19.
[193] Luo, D.C., Zhang, G.X., Liu, J.F., and Sun, X.M. Journal of Physical Chemistry C, 2011, 115, 11327.
[194] Zhao, G., Li, J., Ren, X., Chen, C., and Wang, X. Environmental Science and Technology, 2011, 45, 10454.
[195] Zhao, G., Ren, X., Gao, X., Tan, X., Li, J., Chen, C., Huang, Y., and Wang, X. Dalton Transactions, 2011, 40, 10945.
[196] Zhao, G., Wen, T., Yang, X., Yang, S., Liao, J., Hu, J., Shao, D., and Wang, X. Dalton Transactions, 2012, 41, 6182.
[197] He, Y.Q., Zhang, N.N., and Wang, X.D. Chinese Chemistry Letters, 2011, 22, 859.
[198] Zhang, N., Qiu, H., Si, Y., Wang, W., and Gao, J. Carbon, 2011, 49, 827.
[199] Liu, L., Li, C., Bao, C., Jia, Q., Xiao, P., Liu, X., and Zhang, Q. Talanta, 2012, 93, 350.
[200] Sun, H., Cao, L., and Lu, L. Nano Research, 2011, 4, 550.
[201] Liu, M., Chen, C., Hu, J., Wu, X., and Wang, X. Journal of Physical Chemistry C, 2011, 115, 25234.
[202] Chandra, V., Park, J., Chun, Y., Lee, J.W., Hwang, I.C., and Kim, K.S. ACS Nano, 2010, 4, 3979.
[203] Wu, X.L., Wang, L., Chen, C.L., Xu, A.W., and Wang, X.K. Journal of Materials Chemistry, 2011, 21, 17353.
[204] Ma, H.L., Zhang, Y., Hu, Q.H., Yan, D., Yu, Z.Z., and Zhai, M. Journal of Materials Chemistry, 2012, 22, 5914.
[205] Zhu, J., Wei, S., Gu, H., Rapole, S.B., Wang, Q., Luo, Z., Haldolaarachchige, N., Young, D.P., and Guo, Z. Environmental Science and Technology, 2012, 46, 977.
[206] Zhang, S., Shao, Y., Liu, J., Aksay, I.A., and Lin, Y. ACS Applied Materials & Interfaces, 2011, 3, 3633.
[207] Liu, F., Chung, S., Oh, G., and Seo, T.S. ACS Applied Materials & Interfaces, 2012, 4, 922.
[208] Ramesha, G.K., Kumara, A.V., Muralidhara, H.B., and Sampath, S. Journal of Colloid and Interface Science, 2011, 361, 270.
[209] Gao, Y., Li, Y., Zhang, L., Huang, H., Hu, J., Shah, S.M., and Su, X. Journal of Colloid and Interface Science, 2012, 368, 540.
[210] Pei, Z., Li, L., Sun, L., Zhang, S., Shan, X.Q., Yang, S., and Wen, B. Carbon, 2013, 51, 156.
[211] Maliyekkal, S.M., Sreeprasad, T.S., Krishnan, D., Kouser, S., Mishra, A.K., Waghmare, U.V., and Pradeep, T. Small, 2013, 9, 273.
[212] Li, J., Zhang, S., Chen, C., Zhao, G., Yang, X., Li, J., and Wang, X. ACS Applied Materials & Interfaces, 2012, 4, 4991.
[213] Yang, S.T., Chang, Y., Wang, H., Liu, G., Chen, S., Wang, Y., Liu, Y., and Cao, A. Journal of Colloid and Interface Science, 2010, 351, 122.
[214] Gao, W., Majumder, M., Alemany, L.B., Narayanan, T.N., Ibarra, M.A., Pradhan, B.K., and Ajayan, P.M. ACS Applied Materials & Interfaces, 2011, 3, 1821,
[215] Madadrang, C.J., Kim, H.Y., Gao, G., Wang, N., Zhu, J., Feng, H., Gorring, M., Kasner, M.L., and Hou, S. ACS Applied Materials & Interfaces, 2012, 4, 1186.
[216] Lee, Y.C., and Yang, J.W. Journal of Industrial & Engineering Chemistry, 2012, 18, 1178.
[217] Huang, Z.H., Zheng, X., Lv, W., Wang, M., Yang, Q.H., and Kang, F. Langmuir, 2011, 27, 7558.
[218] Deng, X., Lue, L., Li, H., and Luo, F. Journal of Hazardous Materials, 2010, 183, 923.

[219] Hao, L., Song, H., Zhang, L., Wan, X., Tang, Y., and Lv, Y. Journal of Colloid and Interface Science, 2012, 369, 381.
[220] Chandra, V., and Kim, K.S. Chemical Communications, 2011, 47, 3942.
[221] Li, Y., Zhang, P., Du, Q., Peng, X., Liu, T., Wang, Z., Xia, Y., Zhang, W., Wang, K., Zhu, H., and Wu, D. Journal of Colloid and Interface Science, 2011, 363, 348.
[222] Bradder, P., Ling, S.K., Wang, S., and Liu, S. Journal of Chemical and Engineering Data, 2011, 56, 138.
[223] Li, Y.H., Liu, T., Du, Q., Sun, J., Xia, Y., Wang, Z., Zhang, W., Wang, K., Zhu, H., and Wu, D. Chemical and Biochemical Engineering Quarterly, 2011, 25, 483.
[224] Sun, L., Yu, H., and Fugetsu, B. Journal of Hazardous Materials, 2012, 203–204, 101.
[225] Sui, Z., Meng, Q., Zhang, X., Ma, R., and Cao, B. Journal of Materials Chemistry, 2012, 22, 8767,
[226] Cheng, J.S., Du, J., and Zhu, W. Carbohydrate Polymers, 2012, 88, 61,
[227] Thuy-Duong, N.P., Viet Hung, P., Kim, E.J., Oh, E.S., Hur, S.H., Chung, J.S., Lee, B., and Shin, E.W. Applied Surface Science, 2012, 258, 4551.
[228] Xie, G., Xi, P., Liu, H., Chen, F., Huang, L., Shi, Y., Hou, F., Zeng, Z., Shao, C., and Wang, J. Journal of Materials Chemistry, 2012, 22, 1033.
[229] Wang, C., Feng, C., Gao, Y., Ma, X., Wu, Q., and Wang, Z. Chemical Engineering Journal, 2011, 173, 92.
[230] Li, N., Zheng, M., Chang, X., Ji, G., Lu, H., Xue, L., Pan, L., and Cao, J. Journal of Solid State Chemistry, 2011, 184, 953.
[231] Bai, S., Shen, X., Zhong, X., Liu, Y., Zhu, G., Xu, X., and Chen, K. Carbon, 2012, 50, 2337.
[232] Wu, T., Cai, X., Tan, S., Li, H., Liu, J., and Yang, W. Chemical Engineering Journal, 2011, 173, 144.
[233] Li, Y., Du, Q., Liu, T., Sun, J., Jiao, Y., Xia, Y., Xia, L., Wang, Z., Zhang, W., Wang, K., Zhu, H., and Wu, D. Materials Research Bulletin, 2012, 47, 1898.
[234] Sen Gupta, S., Sreeprasad, T.S., Maliyekkal, S.M., Das, S.K., and Pradeep, T. ACS Applied Materials & Interfaces, 2012, 4, 4156.
[235] Zhao, G., Li, J., and Wang, X. Chemical Engineering Journal, 2011, 173, 185.
[236] Hu, R., Dai, S., Shaoa, D., Alsaedi, A., Ahmad, B., and Wang, X.K. Journal of Molecular Liquids, 2015, 203, 80.
[237] Meyer, Jannik C., Geim, A. K., Katsnelson, M. I., Novoselov, K. S., Booth, T. J., and Roth, S. Nature, 2007, 446, 60.
[238] Leenaerts, O., Partoens, B., and Peeters, F. M. Applied Physics Letters, 2008, 93(19), 3107.
[239] Bunch, J. S., Verbridge, S. S., Alden, J. S., van der Zande, A. M., Parpia, J. M., Craighead, H. G., and McEuen, P. L. Nano Letters, 2008, 8, 2458.
[240] Chen, S., Brown, L., Levendorf, M., Cai, W., Ju, S.Y., Edgeworth, J., Li, X., Magnuson, C. W., Velamakanni, A., Piner, R.D., Kang, J., Park, J., and Ruoff, R. S. ACS Nano, 2011, 5, 1321.
[241] Yuan, W., Chen, Ji., and Shi, G. Materials Today, 2014, 17, 2.
[242] Fischbein, M.D., and Drndic, M. Applied Physics Letters, 2008, 93(11), 3107.
[243] Bai, J., Zhong, X., Jiang, S., Huang, Y., and Duan, X. Nature Nanotechnology, 2010, 5, 190.
[244] Safron, N. S., Brewer, A. S., and Arnold, M. S. Small, 2011, 7, 492.
[245] Safron, N. S., Kim, M., Gopalan, P., and Arnold, M. S. Advanced Materials, 2012, 24, 1041.
[246] Akhavan, O. ACS Nano, 2010, 4, 4174.
[247] Liu, J., Cai, H., Yu, X., Zhang, K., Li, X., Li, J., Pan, N., Shi, Q., Luo, Y., and Wang, X. Journal of Physical Chemistry C, 2012, 116, 15741.
[248] Han, T.H., Huang, Y.K., Tan, A.T.L., Dravid, V.P., and Huang, J. Journal of American Chemical Society, 2011, 133(39), 15264.
[249] Sun, C., Boutilier, M.S.H., Au, H., Poesio, P., Bai, B., Karnik, R., and Hadjiconstantinou, N.G. Langmuir, 2014, 30, 675.

[250] Du, H., Li, J., Zhang, J., Su, G., Li, X., and Zhao, Y. Journal of Physical Chemistry C, 2011, 115, 23261.
[251] Koenig, S.P., Wang, L., Pellegrino, J., and Bunch, J.S. Nature Nanotechnology, 2012, 7, 728.
[252] Konatham, D., Yu, J., Ho, T.A., and Striolo, A. Langmuir, 2013, 29, 11884.
[253] Cohen-Tanugi, D., and Grossman, J.C. Nano letters, 2012, 12, 3602.
[254] Sun, P., Zhu, M., Wang, K., Zhong, M., Wei, J., Wu, D., Xu, Z., and Zhu, H. ACS Nano, 2013, 7, 428.
[255] Sint, K., and Wang, B. Journal of American Chemical Society, 2008, 130, 16448.
[256] Suk, M.E., and Aluru, N.R. Journal of Physical Chemistry Letters, 2010, 1, 1590.
[257] Konkena, B., and Vasudevan, S. Journal of Physical Chemistry Letters, 2012 3, 867.
[258] Choi, E.Y., Han, T.H., Hong, J., Kim, J.E., Lee, S.H., Kim, H.W., and Kim, S.O. Journal of Materials Chemistry,2010(20), 1907.
[259] Paul, D.R. Science, 2012, 413, 335.
[260] Hu, M., and Mi, B. Environmental Science and Technology, 2013, 47, 3715.
[261] Joshi, R.K., Carbone, P., Wang, F.C., Kravets, Y. Su., Grigorieva, I.V., Wu, H.A., Geim, A.K., and Nair, R.R. Science, 2014, 343, 752.
[262] O'Hern, S.C., Boutilier, M.S.H., Idrobo, J.C., Song, Y., Kong, J., Laoui, T., Atieh, M., and Karnik, R. Nano letters, 2014, 14, 1234.
[263] Akin, I., Zor, E., Bingol, H., and Ersoz, M. Journal of Physical Chemistry B, 2014, 118, 5707.
[264] Akhavan, O., and Ghaderi, E. ACS Nano, 2010, 4, 5731.
[265] Perreault, F., Tousley, M.E., and Elimelech, M. Environmental Science and Technology Letters, 2014, 1, 71.
[266] Choi, W., Choi, J., Bang, J., and Lee, J. ACS Applied Materials & Interfaces, 2013, 5, 12510.
[267] Hua, X., Yub, Y., Zhou, J., Wang, Y., Liang, J., Zhang, X., Chang, Q., and Song, L. Journal of Membrane Science, 2015, 476, 200.
[268] Xue, M., Qiu, H., and Gui, W. Nanotechnology, 2013, 24, 505720.
[269] Zhang, H., Lv, X., Li, Y., Wang, Y., and Li, J. ACS Nano, 2010, 4, 380.
[270] Li, N., Liu, G., Zhen, C., Li, F., Zhang, L., and Cheng, H.M. Advanced Functional Materials, 2011, 21, 1717.
[271] Perera, S.D., Mariano, R.G., Vu, K., Nour, N., Seitz, O., Chabal, Y., and Balkus, K.J. Jr ACS Catalysis, 2012, 2, 949.
[272] Yang, N., Liu, Y., Wen, H., Tang, Z., Zhao, H., Li, Y., and Wang, D. ACS Nano, 2013, 7, 1504.
[273] Ghasemi, S., Setayesh, S.R., Habibi, A.Y., Hormozi, M.R.N., and Gholami, M.R. Journal of Hazardous Materials, 2012, 199–200, 170.
[274] Chen, F.J., Cao, Y.L., and Jia, D.Z. CrystEngComm, 2013, 15, 4747.
[275] Lü, W., Chen, J., Wu, Y., Duan, L., Yang, Y., and Ge, X. Nanoscale Research Letters, 2014, 9, 148.
[276] Fu, D., Han, G., Chang, Y., and Dong, J. Materials Chemistry and Physics, 2012, 132, 673.
[277] Xu, T.G., Zhang, L.W., Cheng, H.Y., and Zhu, Y. Applied Catalysis, B: Environmental, 2011, 101, 382.
[278] Li, B., and Cao, H. Journal of Materials Chemistry, 2011, 21, 3346.
[279] Li, Q., Guo, B., Yu, J., Ran, J., Zhang, B., Yan, H., and Gong, J.R. Journal of American Chemical Society, 2011, 133(28), 10878.
[280] Yao, Y., Yang, Z., Zhang, D., Peng, W., Sun, H., and Wang, S. Industrial & Engineering Chemistry Research, 2012, 51, 6044.
[281] Fu, Y.S., Xiong, P., Chen, H.Q., Sun, X., and Wang, X. Industrial & Engineering Chemistry Research, 2012, 51, 725.
[282] Fu, Y., Chen, H., Sun, X., and Wang, X. Applied Catalysis B: Environmental, 2012, 111–112, 280.

[283] Liu, S., Tian, J.Q., Wang, L., Luo, Y.L., and Sun, X.P. Catalysis Science and Technology, 2012, 2, 339.
[284] Gao, Z.Y., Liu, J.L., Xu, F., Wu, D.P., Wu, Z.L., and Jiang, K. Solid State Sciences, 2012, 14, 276.
[285] Chandra, S., Das, P., Bag, S., Bharb, R., and Pramanik, P. Materials Science and Engineering B, 2012, 177, 855.
[286] Zhu, M.S., Chen, P.L., and Liu, M.H. ACS Nano, 2011, 5, 4529.
[287] Tu, X., Luo, S., Chen, G., and Li, J. Chemistry- A European Journal, 2012, 18, 14359–14366.

Farheen Khan and Amin Fathi Amin Ajlouni

4 Characterization of eco-friendly bentonite materials and their applications

4.1 Introduction

Bentonite materials are small particles from dirt, ceramic, filler materials, and so on that are present in deep geological area in the form of clay and clay minerals such as montmorillonite [1]. The chemical inertness and inorganic and nonirritating reformation has not been established for such organic materials. The bentonite materials are mainly originated through weathering of volcanic ash, radioactive wastes, hard rock grains with high iron and swelling content mineral, which indicates existence of water contents in constituents. The compacted soil liners, geosynthetic clays, and composites with geomembranes have already been utilized as different barrier layers in various applications [2]. On the basis of filler materials, typical mineral layers are used as filler content, which shows high degree of filling in the range of 2–4 wt%. Adsorption of water than fifteen times water creates swelling, and their valuable characteristic properties are used for the suspension or colloidal solution in thickness of liquids. The large quantities of bentonite materials are also used in gas and oil drilling industries to suspend high specific gravity slurries and float out lumps of stun amended by the little drill [3]. Particularly, suspension/colloidal solutions and industrial products coated with ceramic materials and reverse electrolytic charges grow on the surfaces and edges of dispersed particles, which rise to stable "house-of-cards" manufacture and disturbed by trim stress. However, when strain is eliminated, the structure reestablishes itself [4]. The maximum overhanging benefit that is achieved by blunging bentonite with water adding other dry materials clearly ensures that all particles are stimulated on the surface of bentonite materials. The bentonite particles bind with ceramic groups to make them stronger in the green or dry state. Its smaller particle size fills spaces and produces compact mass with more points of interaction and also formed by dry strength and harder and more durable surfaces. The hard and strong bentonite is formed by mixing silica and plastic in the ratio of 25:75 [5]. The electrolytic behavior of bentonite particles, shapes, sizes, and surfaces exhibiting high plasticity but excess amounts of bentonite materials show low drying rates. The drying performance of bentonite materials makes it more plastic, harder, and dry, which is the main reason behind cracking behavior of these materials. The prepared plastic materials are mixed together in ratio of 20–30% virgin materials with calcined materials, and specimen test performance has measured shrinkage rates, drying

Farheen Khan, Amin Fathi Amin Ajlouni, Department of Chemistry, Taibah University, Madina, Yanbu, Saudi Arabia

https://doi.org/10.1515/9783110650600-004

behaviors, and mechanical strengths. Bentonite materials act as plastic and are more harder at drying conditions. But dry ingredient blending is only possible by mixing with a high powerful speed propeller [6]. The soluble salts were gray to deep red colored, which is the main reason of firing and utility of its standard-grade granular iron materials, and microfine grades are more significant speck-causing particles makes clay bodies' rated at 325 mesh materials. Association of iron content is good for materials that are more expensive and valuable. However, added barium carbonate precipitate produced white body bentonite materials, which can bear up to 5% bentonite without firing significantly, as compared to darker materials. The white firing bentonite materials are highly refined materials and they exhibit less plasticity. The white firing bentonite raw material contained 5% of microfine materials and 0.5% of iron but kaolin and ball clay contained more iron than the bentonite. If the percentage of iron content increases in bentonite materials, then they are not considered as white plasticizers [7]. The penetration of water in bentonite materials slowly moves but not reaches the deep areas, resulting in bentonite-containing clay body dry slower, tightly bound and water remains in some part of materials. The dried bentonite materials are affected by high temperature, and the high steam pressure generates fracture or cracking which is risky for fire. On the other hand, the compacted bentonite materials restrict the migration of water including varieties of contaminants while pure bentonite has the tendency to pelletization showing high plasticity and low heat conductivity in water soil process, and it hinders the blocking ability of the buffer/backfill materials. Many researchers have described the characteristic properties of bentonite and bentonite mixtures such as grain size, grain density, specific surface area, permeability, chemical activity, strength swellling pressure, hydraulic conductivity, and mineralogical and sealing properties in detail. The most relevant properties with respect to bentonite have been used for repository and other potential buffer and tunnel backfilling materials [8]. The mineralogical variations and sealing properties are the alternative sources of bentonite, which quantify various types of composite bentonite materials consequently valuable in order to secure quality, supply, and prices. The important aspects of compacted bentonite materials are availability, existences in short and long term in groundwater sympathy, hazardous substances present in regular manner, price, and so on. Most of the analyzed materials contain more materials and denote huge clay formations, and their ingredients are used as ceramics, waterproofing, sealing in landfill sites, nuclear waste repositories, filter, and stabilizer for adhesives, paints, cosmetics, and medicines [9–14]. Determination of mineral composition in compacted and clay materials is an unconventional analysis; therefore, many techniques have been employed for homoionic dialyzed clay fraction element analysis in bentonite bulk material. Techniques such as inductively coupled plasma-atomic emission spectrometry, carbon analyzer (RC412) for carbon element and sulfur for combustion technique, mineralogical composition for X-ray diffraction (XRD), cationic exchange capacity (CES), energy-dispersive spectroscopy, thermogravimetry (TG), differential scanning calorimetry, Fourier transform infrared spectroscopy (FTIR), helium picnometry, and scanning electron microscopy (SEM) [15] have been employed. Applications of these materials including clays show considerable attention in various fields: adsorbents for removal of heavy metals from water and effluents, pollution control, heat consumption in modern economies (transportation process), fossil fuels, biofuels, catalysts, and nanocomposite polymer as nanofiller, spherical or cubical nanofillers based on the nanometer range (nanosilica, carbon black, nano-$CaCO_3$, etc.), rod- or fiber-type nanofillers (carbon nanotubes and cellulose whiskers), and sheet-like nanofillers (clay and layered silicates). The characterization results of bentonite materials are promising and better understood as they are considered for environmental utilizations [16–19].

This chapter describes the specific accessory minerals that exist in different forms of bentonite materials, their remarkable characterization properties and their significant effect on chemical and physical parameters under the influence of nature. The possible applications of bentonite materials are evident in enormous fields, and they are used as catalysts, adsorbents, composite nanomaterials, and nanofillers.

4.2 Chemical composition of materials

The mixing of two or more distinct products consists of complicated materials, which have significantly different physical and chemical properties that vary from individual components. The chemical composition identifies component of content present in materials.

4.2.1 Bentonite

Bentonite materials exhibit different chemical compositions with various physical and chemical properties. Compact-based bentonite materials include potassium (K), sodium (Na), calcium (Ca), aluminum (Al) elements along with Mg, Fe, Li, but Na and Ca, which are the major constituents of bentonite materials (clay and montmorillonite). The alteration of clay glassy mixture produced from sodium montmorillonite and calcium montmorillonite with large deposition of clay and montmorillonite makes bentonite. The general formula of bentonite is Al_2O_{34} SiO_2. H_2O but the word bentonite is mainly coined by mixing of biggest smectite groups (clay and montmorillonite) [20, 21].

4.2.2 Smectite group

Smectite group (clay minerals) is the part of bentonite materials comprising minerals such as montmorillonite, beidellite, nontronite, hectrorite, and saponite. The chemical formulas of each member of smectite group are given in Table 4.1 [11].

Table 4.1: Chemical composition of bentonite materials (smectite groups).

S. no.	Mineral names	Chemical composition
1	Montmorillonite	$M^{+}_{0.33}$ $(Al_{1.67}$ $Mg_{0.33})$ Si_4O_{10} $(OH)_2.nH_2O$
2	Beidellite	$M^{+}_{0.33}$ $Al_2(Si_{0.367}$ $Al_{0.33})$ O_{10} $(OH)_2.nH_2O$
3	Nontronite	$M^{+}_{0.33}$ Fe_2^{3+} $(Si_{0.367}$ $Al_{0.33})$ O_{10} $(OH)_2.nH_2O$
4	Hectrorite	$M^{+}_{0.33}$ $(Mg_{0.267}$ $Li_{0.33})$ Si_4O_{10} $(OH)_2.nH_2O$
5	Saponite	$M^{+}_{0.33}$ $(Mg_{0.267}$ $R^{3+}_{0.33})$ $(Si_{0.334}$ $Al_{0.66})$ O_{10} $(OH)_2.nH_2O$

Source: Evans, Jeffrey C. "Vertical cutoff walls." Geotechnical practice for waste disposal. Springer, Boston, MA, 1993. 430–454.

Here M^+ represents Na^+, K^+, and Ca^{2+} cations, R^{3+} represents Al^{3+}, Fe^{3+}, Cr^{3+} trivalent cations, and xH_2O represents water molecules present in smectite groups [22, 23].

4.2.3 Clay

Basically the general formula of clay (containing smectite group) is (Ca, Na, H)(Al, Mg, Fe, Zn)$_2$(Si, Al)$_4O_{10}(OH)_2.xH_2O$. But clay materials are mainly comprised of kaolinite, montmorillonite, vermiculite, and chlorite. The mineral aggregate units assure alternating sheet of SiO_2 and AlO_6. The units of SiO_2 and AlO_6 ratio are present in the ratio of 1:1 kaolinite, 2:1 montmorillonite, 2:1 vermiculite, and 2:2 chlorites, which are capable of holding water molecules. The oxide negative charge neutralizes by K^+, Ca^{2+}, and Mg^{2+} cations [24].

4.2.4 Montmorillonite

The mixing of quartz grains and crystals in sandstone formed bright pink chalky massive materials, clay-like fracture, and invisible crystalline affinities. The general formula of montmorillonite is same as clay (contain smectite group) (Na, Ca)(Al, Mg)$_6$ $(Si_4O_{10})_3(O\ H)_6{-}nH_2O$ and the presence of sodium element shows variance from clay materials. The most important montmorillonite minerals are sodium and calcium montmorillonite used in large-scale industries, which show different physicochemical properties [25].

4.3 Materials structure

4.3.1 Structure of bentonite

Bentonite structures mainly consist of aluminum phyllosilicate with impure clay and montmorillonite minerals. These structures are based on two structural units such as an octahedral layer and tetrahedral layer.

4.3.1.1 Octahedral layer

In three-layer sandwich pallets, octahedral layer consists of aluminum (Al_2O_3) layer that exists in the center of the body pallets and tetrahedral layer (SiO_2) outside the pallets. The octahedral structure makes oxygen or hydroxyl cations, which orients in the middle position at the same distance. The octahedral structure is shown in Figure 4.1.

(a) (b) (c)

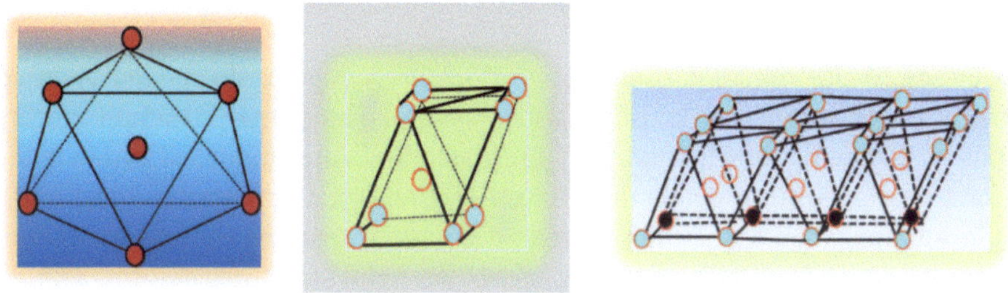

Figure 4.1: (a) Octahedral structure, (b) single octahedral unit, and (c) octahedral sheet structure.

In the octahedral structure, if Al(III) present in the center occupies two-thirds of positions, then it is called gibbsite, and if Mg(II) is present, then it is called brusite. The formula of gibbsite is $Al_2(OH)_3$ and brusite is $Mg_3(OH)_6$ cations. These metal ions (Ca^{2+}, Mg^{2+}, and Na^+) present in bentonite structure shows ion–dipole forces with water molecules and migrates toward the outside silica layers causing hydration in crystal lattice.

4.3.1.2 Tetrahedral layer

Tetrahedral layer consists of tetrahedral oxygen or hydroxyl atoms arranged in all four corners and Si atom is sited at the center of tetrahedron, which is shown in Figure 4.2. The oxygen or hydroxyl atoms contribute much to silicon in the tetrahedral

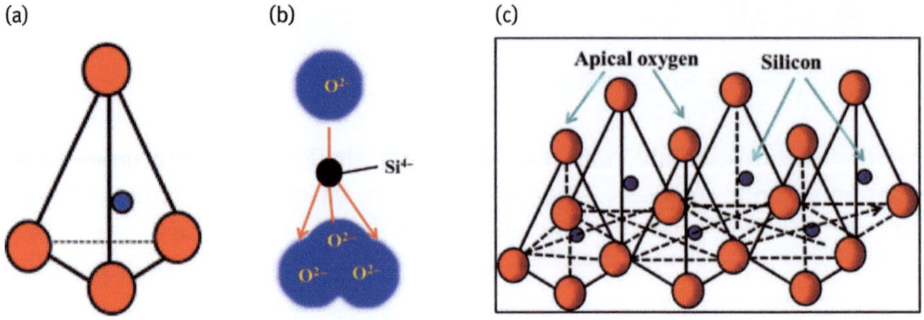

Figure 4.2: (a) Tetrahedral structure; (b) single octahedral unit; and (c) tetrahedral sheet structure.

layer to form hexagonal-like structure and base for all face same direction in the same plane. The octahedral and tetrahedral layers associated with shared oxygen or hydroxyl atoms produced different compositions of mineral materials such as clay minerals and montmorillonite [26–28].

4.3.2 Structure of clay

The structure of clay minerals consists of cations prescribed in sheets and oriented in tetrahedral or octahedral manner (with oxygen). The planar structure of clay minerals raises the platelet habitat, and alternated units are supported by materials in 2:1 or 1:1 tetrahedral or octahedral ratio. Additionally, clay minerals have interlayers located between sequential 2:1 units occupied by cations causing hydration. The pressure applied on hydrated or moistened material structures deforms and removes clay to retain their original shapes. The tendency of natural layer (1:1 or 2:1) dioctahedral or trioctahedral character are based on material classification in octahedral sheet (Figure 4.3) [29, 30].

4.3.3 Structure of montmorillonite

Sodium and calcium montmorillonite are natural, desirable, fine-grained bentonite materials, which are composed of clay minerals. Chemically and structurally, montmorillonites are described as hydrous aluminum silicates, which are very stable molecules and they contain small amount of alkali and alkaline earth metals. Basically, it consists of aluminum octahedral sheet and silica tetrahedral sheet. A single unit cell consists of two silica tetrahedral sheets in between aluminum octahedral sheet as shown in Figure 4.4.

The montmorillonite lattice is negatively charged and balanced by cations on the surface of fragments. Cations are alleged in this manner by the clay minerals

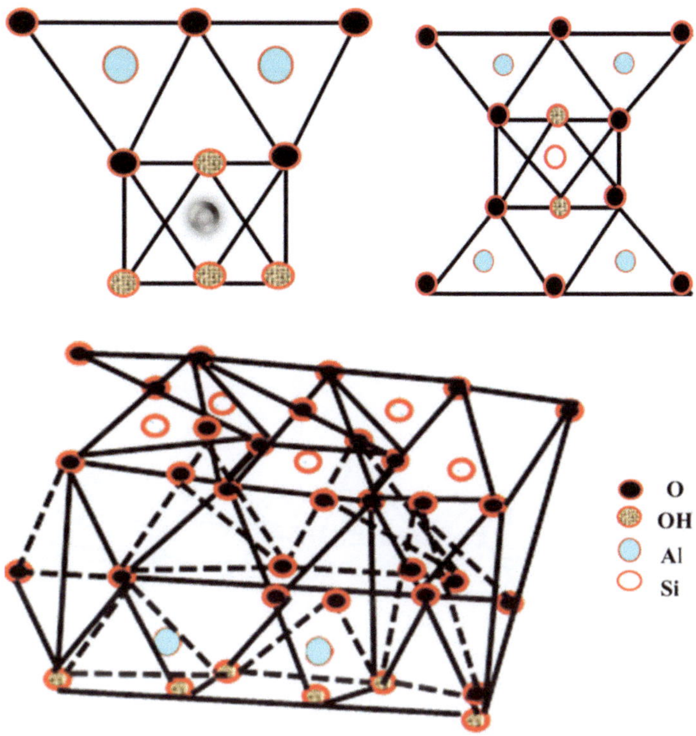

Figure 4.3: Clay mineral sheets composed of dioctahedral and trioctahedral structural units.

that freely exchange cations indicated by CEC, measured in milliequivalents/100 g (meq/100 g) [31]. The cations most commonly found in nature are sodium and calcium (Figure 4.5). The main feature of the montmorillonite structure is water or other polar molecules that enter in between the unit layers and expand in the vertical direction. The prominent cations and anions (Ca^{2+}, Mg^{2+}, H^{+}, K^{+}, NH^{4+}, Na^{+}, SO_4^{2-}, Cl^{-}, PO_4^{2-}, and NO_3^{2-}) are found on clay surface. These ions are exchanged with other ions relatively easily without affecting the clay mineral structures [32, 33].

4.4 Behavior and characterizations

The behavior of bentonite materials is very complex and is described by models and concepts such as Barcelona expansive model (BExM), double structure models (DBM), hydraulic conductivity, thermal conductivity permeability, swelling capability, plasticity, high chemical buffering capacity, stiffness, grain size, porosity, density, and surface area. The study focused on behavior and physicochemical properties, which described structure, morphology, and characterizations of bentonite materials such

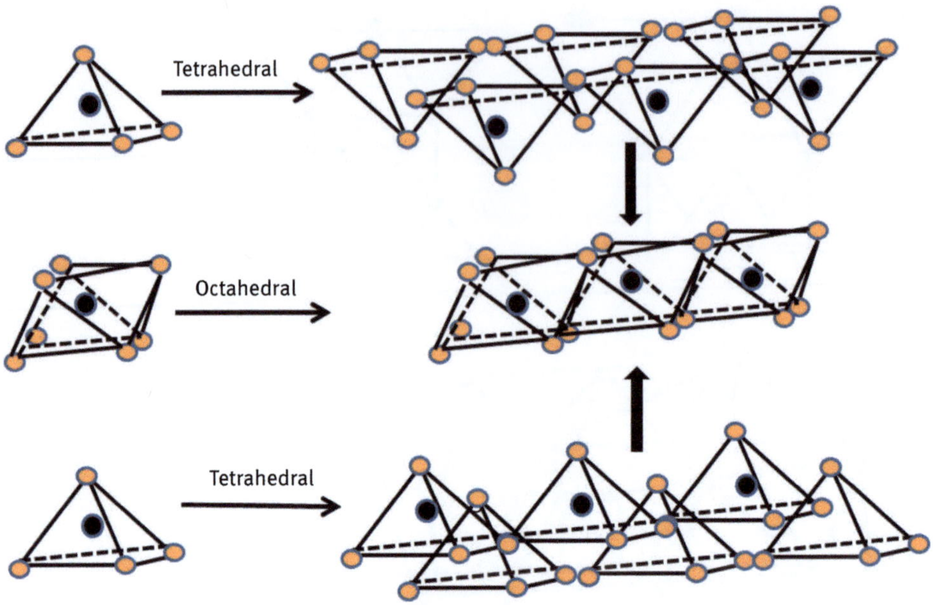

Figure 4.4: Structure of montmorillonite.

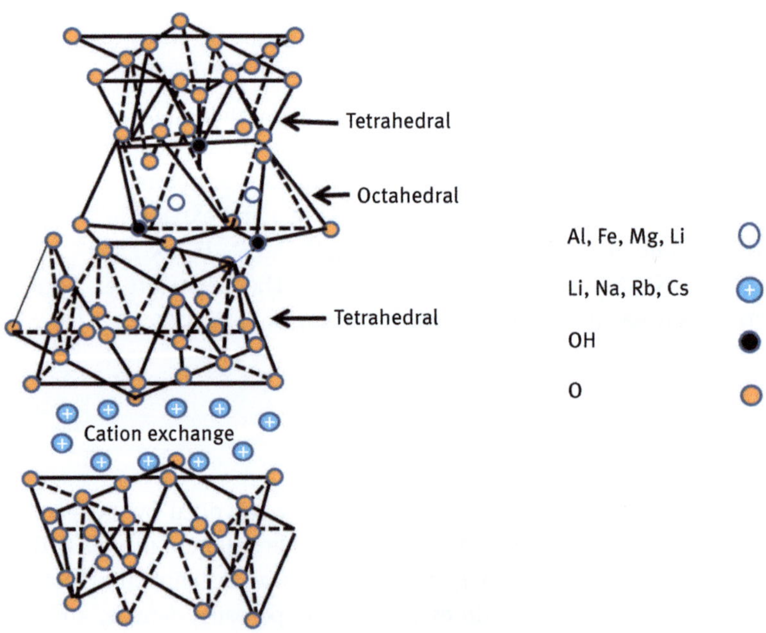

Figure 4.5: Exchange of cations from montmorillonite.

as clay soil. The small pore size particles are connected to distinct constituents and diffused into two structure layers [34–38].

4.4.1 Models

A number of constitutive models have conducted the behavior of compacted bentonite materials. With regard to experimental results, many authors have suggested different models to better understand the thermohydromechanical behavior of bentonite materials. Gens and Alonso [39] proposed the elastoplastic model for expansive clay materials, considered water potential in two structure layers, and confirmed basic pore levels by the experimental method. In these models, the main aspects highlighted are macro- and microstructure levels and their interaction between two layer structures with irreversible behavior of applied pressure and difficulty to determine its initial yields. The fact of this model framework of elastoplasticity and explicitly theory yielded function that was not defined because it doesn't give clear evidences for shape and internal yielding surface interaction mechanism between two structure layers. Alanso et al. [40] reported numeric data of swelling materials, and the recommended BExM showed stress variables of expensive materials in unsaturated conditions. Sanchez et al. [41] extended the above consecutive models and give mathematical framework DBM elastoplastic for strain hardened materials [40, 41]. With respect to DBM described soil behavior and several mechanisms interaction combined together or simultaneously. Cui et al. [42] reported a unique relationship between swelling pressure and void ratios, which is found in the structure of clay soil, produced high pressure and absorbed more amount of water with high hydraulic conductivity. The voids located in the interfaces of material surface show greater effect on the ratio of clay soil and volume of water. A series of parameters such as water retention test, hydration test, suction-controlled oedometer test, and hydraulic test have been performed on bentonite–sand mixture samples for the analysis of voids [42].

4.4.2 Physical and chemical properties

The physical and chemical properties of bentonite materials are described in below.

4.4.2.1 Grain size

The substantial materials of bentonite and clay are commonly obtained from natural earth rock. The small-sized grain particles existing in less amount of water indicate plastic behavior. For chemical analysis, substantial materials mainly comprised silicon (Si), aluminum (Al), water (H_2O), iron (Fe), alkali metals (Na), and alkaline earth metals (Mg) and categorized fraction particles on the basis of smaller sizes. The size of material particles varies with different disciplines, more than 4 µm is the maximum grain size and less than 2 µm is the minimum size but more than 2 µm particle size is present in clay soil materials [43]. Large- and small-sized dry grain particles are accumulated on dry conditions. Some dry grain particles are deformed and dispersed in water and air produced dust problems. The high-resolution particle size distribution is confirmed by XRD, and gravitational and disk centrifugal sedimentation techniques. When compare with nonclay materials, the sizes of grain particles are larger than clay particles and minerals as below 5%. Additionally, clay minerals and clay-grade particles make up less than half of rock but in some cases nonclay particles have maximum size of the clay grade particles, although certain materials consist of moistened or plastic, which does not contain coarse-grained material and placed in silt or sand category [44].

4.4.2.2 Grain density

The percentage of voids space present in rock materials is called porosity and it depends on the grain size of particles present in discipline manner varies from 1% to 40%. For example: crystalline structure of granite has low porosity (1%) and sandstone shows higher porosity (10–35%) as sand or mineral grains does not fit together and produced larger pore space in mineral materials. The term density is defined as mass per volume; porosity filled by fluid with dry or grain size particles also known as wet density, dry density, and grain density. The pycnometer was employed for the measurement of material density, and specific grain density was recorded in the range of ±10 kg/m^3 [45].

4.4.2.3 Specific surface area and charges

The Brunauer–Emmett–Teller (BET) technology on micromeritics ASAP 2400 instrument was determined by the specific surface area of mineral materials under degassed vacuum conditions weighed and cooled by liquid nitrogen. It is a very important parameter, which quantified dissolution of minerals and interaction of particles in soil and sediments. Atomic force microscopy, N_2 gas adsorption (BET), and liquid adsorption used ethylene glycol monomethyl ether techniques were used to analyze the specific surface area of clay materials. As compared to measurement technique, AFM specific surface area was greater than the respective gas adsorption (N_2-BET) specific surface area. The stability of colloidal particles is related to surface area with surface charges connected to zeta potential in the form of electric charge diffused into layer of colloidal solutions and carry negative charge dependent on ionic strength and its pH [46–55].

4.4.2.4 Swelling

Swelling of bentonite materials shows favorable properties, which donated passage of water molecules inside the structure of mineral layers. Its functional property is considered as the key of backfill materials. Many authors have reported the swelling properties of bentonite and clay–sand mixtures as a bundled plate structures and its surface is occupied by sodium ions. These ions are present in water solution with decrease in water chemical potential and concentration. The differences of water potential and concentration are responsible for easy passages of water molecule or water transport. The transportations of water molecules occur from high potential level to low potential level or low concentration to high concentration levels. The chemical potentials and concentration are equal and lead to uniform ions at equilibrium stages. The regular accumulation of water molecules causes hydration of sodium ions, generating negative charges on the surface of bentonite plate (Figure 4.6). After developed surface electric charges resist each other, platelets are compensated and moved apart from original positions, which is called swelling. The rate of swelling depends on the quality and grades [56–58].

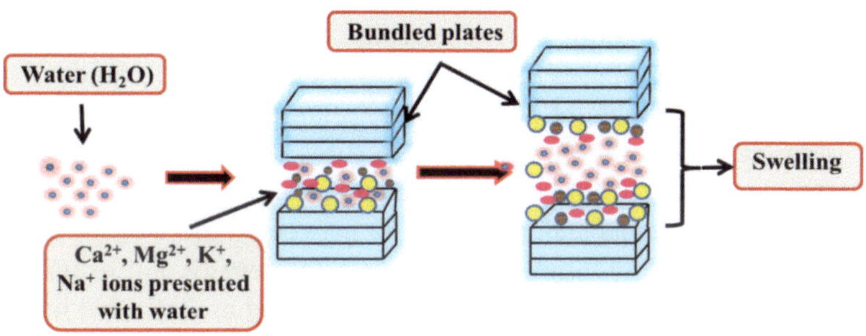

Figure 4.6: Swelling process in bentonite materials.

The powered materials of bentonite grades absorb water gradually, especially grain-sized bentonite particles absorb more water and poured on them. The materials absorbed water is nearly five times the weight of water molecules and occupies a volume of 12 to 15 times its dry bulk weight. In case of montmorillonite, water molecules are intercalated between individual structure layers and formed interlayer of ionic solutions. Mostly cations are constraint diffused in the mineral surface and demanded neutrality of electrical charges. For more swelling, high chemical water potential increases the interlayer distance and transported into the interlayer space of montmorillonite (Figure 4.7). Actually, swelling is a reversible process, which shrinks volume to initiated to its drying and swelling process. For example clay: a number of independently rich water molecules involved in shrink and drying swelling process and entire mass swelled throughout water molecules [59–63].

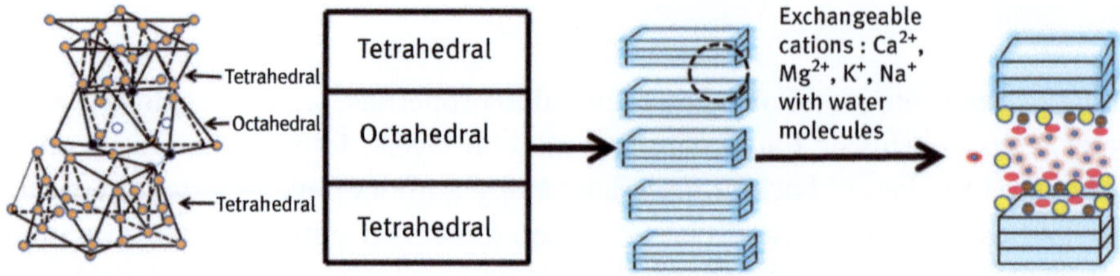

Figure 4.7: Transportations of cations with water molecules in swelling process.

4.4.2.5 Thermal and hydraulic conductivity

The thermal conductivity of compacted bentonite is related to the temperature of wastes or rocks, which distributed heat and transfer that into containers of host rocks under the influence of geological repository. The temperature distribution shows grater effects on mechanical and hydraulic behaviors of barrier system present in other variables such as thermal expansion, phase exchange, and thermal osmosis. The thermal conductivity of the porous buffer materials with the sequence of multiphase fluids in parallel series such as water and air–vapor mixture connected to pore system. It is denoted as a function of porosity, degree of saturation, temperature, and pressures of fluid phases in the barrier system. Hydraulic conductivity was calculated from the volume of percolated solution regarding Darcy's law and measured series in parallel with swelling pressures at equilibrium stages. According to the series of sample analysis, some samples exposed low density in high saline solution conditions; its major differences between mineralogical solutions because they are frequently minor at the same total density. For example, the rheological properties of buffer materials are compensated and affected by an increase in the total density [64–66].

4.5 Purification of bentonite materials

The crude bentonite materials associated with quartz, muscovite, illite, feldspar iron oxide, and other composites are entirely connected to mineral impurities suspended in water and presented in liquid precipitated form (slag) as shown in Figure 4.8.

The mineral impurities of liquid suspended solution are separated from each other and purified by recycling and filtering or centrifuging processes. To separate impurities in suspension, the solution required more water, time, and chemical reagents. Particularly, all ingredients other than montmorillonite are removed from crude bentonite by a slurry solution of crude mineral (sodium polymetaphosphate) with diluted water in more amounts. The solid mineral materials are still settled from separated supernatant liquid and settled precipitate (impurity) completely set

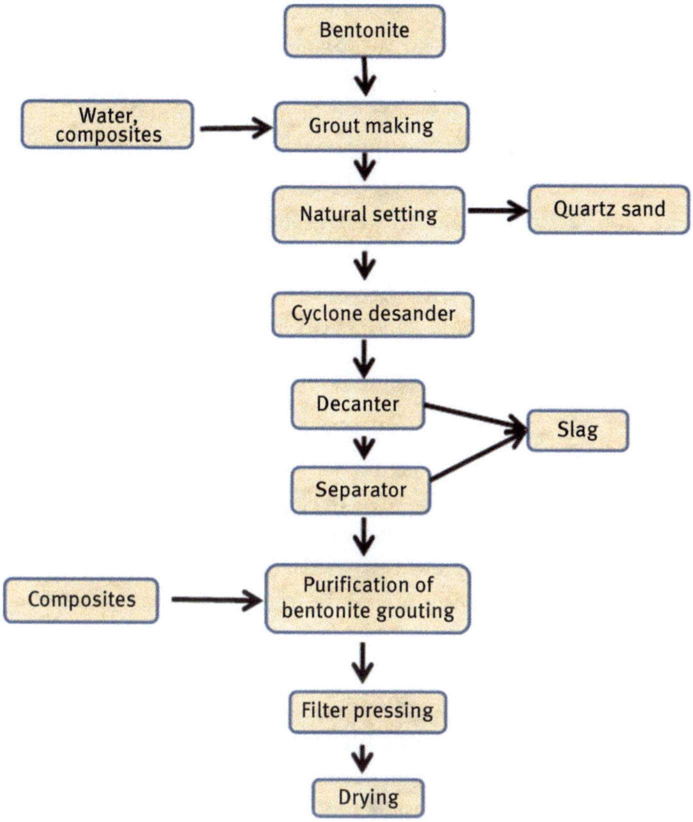

Figure 4.8: Purification process of bentonite materials.

and obtained pure form of montmorillonite. Surprisingly, uniform purity of clay and montmorillonite is achieved from raw materials that vary greatly in nature and they are used in food industry or pharmaceutical industry. Instead, more samples collected from liquid of bentonite materials are applied to filtering and centrifuging techniques. In these techniques, the desired product remains constant with significant amounts of sodium polymetaphosphate and recycled with extra amounts of slurry. The moistened bentonite is dried, powdered, and its powder is further used in this process [67, 68].

4.5.1 Applications

In the last few years, widely used bentonite materials at large levels pronounced structural, mechanical, and thermal properties, hydration, swelling, water absorption, viscosity, thixotropy, ability to act as a bonding agent, stability, high specific surface area with high CEC. In general, feldspar, biotite, kaolinite, illite, cristobalite, pyroxene, zircon, smectite, and chlorite contained clay and montmorillonite minerals [69–72] and other clay-sized crystal minerals such as quartz, carbonate, and metal oxides. The broad application areas of bentonite materials used as organic and inorganic filler materials show significant effects in educational, construction, civil engineering, and industrial fields. Clay materials and its composite materials are involved in various purposes such as adsorbents, purification of water, and removal of organic, inorganic, biological, and water pollutants. Other specific applications are they are used in paints, coatings, polymers as functional filler, cosmetic, pharmaceuticals, and medical. Its applications are described as follows [73].

4.5.2 In industry

On the basis of materials importance and economical requirements, USA is the largest producer with 29% share of the world production in recent years and Turkey is the most demanded country. The bentonite material was used in industries because it has capability to absorb bonding with other waste materials. The preparation of molding sand materials employed in production of iron, steel, and nonferrous casting clumps, after removing the clumps, remaining product further used in mica, talc, and wollastonite minerals are involved for the purpose of industrial coating, anticorrosive, marine coatings, and improved material abilities such as film integrity, water transmission, and reduced water vapor permeability. These minerals provide corrosion resistance, cracking, weathering, and reinforcement for coating applications [74, 75].

4.5.3 In construction and civil engineering

Traditionally, clay from bentonite materials are used for sealing/filler purpose, for example, Portland cement and mortars, which support in building construction and engineering project. The nature of foundation materials are lubricant, viscos, plastic, highly unstable, and unconsolidated material collapse and placed in horizontal directions for drilling and jack piping [76].

4.5.4 Adsorbents used for impurities removal

The natural clay and their composite materials work as adsorbents for removal of organic and inorganic impurity contaminated by pathogens from water circulation in drinking water. The clay adsorbent materials show low permeability, high surface area per unit weight, and high CEC, causing physical and chemical barriers for separation of various types of wastes to avoid toxics from dispersive particles fused in environmental pollutions. On the basis of contaminant impurities, they are divided into four categories and impurities removed by the clay and bentonite materials [77–85].

(a) Heavy metal impurities
(b) Inorganic impurities
(c) Organic impurities
(d) Pathogens

4.5.4.1 Heavy metal impurities

Bentonite materials are good adsorbents to adsorb heavy metal ions. Regular disposal of heavy metals in environments creates health hazardous problems, and lead toxicity has emerged as an important global problem with public health consequences due to its serious impact on brain function, which are mainly connected to water and industrial wastewater for treatment of toxic elements, bonded through clay materials and adsorbed on the surface of adsorbent materials. Therefore, for environmental safety and human health demanded elimination of toxic elements in natural drinking water. In general, water pollution is caused by toxic metal ions from various industrial effluents. The effective concentrations of heavy metal ions {chromium(III)/ (VI), cadmium (II), lead(II), mercury(II), nickel(II) and copper(II)} are mainly referred to waste stream processes such as mining processes, metal plating, power generating, electronic device, manufacturing, and tanneries released over limits. Bentonite, clay, and their modified forms, kaolinite and montmorillonite, have drained much attention to scientists because they exhibit high efficiency for removal of metal ions from aqueous medium, easy availability, and less costly. Montmorillonite and its modified forms have much higher metal adsorption capacity compared to kaolinite and its modified forms. The adsorption of metal ions like Cd, Cr, As, Mn, Fe, Pb, Cu, and Zn by modified form of kaolinite and montmorillonite indicates that clay materials show higher metal adsorbing capacity as compared to their counterparts [86–98] (Table 4.2).

4.5.4.2 Inorganic impurities

Inorganic contaminants such as fluoride and nitrate present in natural drinking water cause methemoglobinemia or blue baby disease. The useful concentration range of fluoride for dental purpose is 1.0–1.5 mg/L, if fluoride found above this concentration range caused harmful anthropogenic factors. For the removal of inorganic contaminants, Zhang et al. [99] successfully applied adsorption technique and prepared low-cost adsorbents (bentonite or chitosan beads). The batch adsorption techniques applied for the sorption of alpha and beta endosulfan onto bentonite clay under the influence of various parameters such as contact time, adsorbent dosage,

Table 4.2: Removal of heavy metal impurities by bentonite clay-modified materials.

S. no.	Impurities of metal ions	Bentonite clay and modified materials	References
1	Cd(II)	Kaolinite, montmorillonite, and Their poly(oxozirconium) and tetrabutyl-ammonium derivatives	[82]
2	As(III)	Montmorillonite, kaolinite, and illite	[83–84]
3	Cu(II), Ni(II), Co(II), Mn(II)	Kaolinite and montmorillonite	[85]
4	Pb(II), Cd(II), Ni(II), Co(II)	Bentonite-methylene bis-acrylamide	[86]
5	Mn(II)	Bentonite-porphyrin carbon paste	[87]
6	Fe(III), Al(III), Pb(II), Cu(II)	Carbonate-modified natural zeolite and mixture of bentonite	[88]
7	Ni(II), Cd(II), Cu(II)	Bentonite clay–iron oxide composite	[89]
8	Pb(II)	Kaolinite and montmorillonite	[90]
9	Zn(II)	Bentonite	[91–92]
10	Pb(II), Zn(II), Ni(II)	Sodium bentonite-activated clay	[93]
11	Pb(II), Zn(II)	Bentonite	[94]
12	U(VI)	Thermally activated bentonite	[95]
13	Cu(II)	Bentonite polyacrylamide organobentonite	[96–97]
14	Ni(II), Co(II)	Na-activated bentonite	[98]

initial endosulfan concentration, pH, and temperature, and evolved equilibrium data by kinetic models, Freundlich isotherm model, Langmuir, Dubinin–Radushkevich (D–R), and Temkin isotherm models followed pseudo-first order and pseudo-second order [100]. The adsorption capacity is of 0.895 mg/g bentonite (3.0 g), chitosan beads is of 0.359 mg/g, and defluoridation capacity is of 1.164 mg/g optimal at 5 pH value, respectively. Phosphorous is one of the major nutrients contributing to increased eutrophication of waterbodies and causes water eminence problems such as increased purification costs, decreased recreational, and conservation value of waterbodies in drinking water. EL-Sergany and Shanableh have synthesized Al-modified bentonite samples for the removal of phosphorous removal. The Al-modified Bentonite materials synthesized by mixing of bentonite with selected Al-solutions are prepared with [OH]/[Al] ratios equal to 2:4. The size of synthesized material ranged between 63 and 425 µm and showed particle size (362, 112, 69 µm) effects on adsorption of phosphate onto Al-PILB. The first- and second-order rate constants K_1 and K_2 are affected by adsorbent size, adsorption capacity increase with decrease in the adsorbent particle size. The study confirmed that high potential of Al-modified clay is used for removal of phosphorus and regulate the particle size of the adsorbent to accomplish phosphorus removal by Yan-kui et al. and Zamparas et al. The Zenith-N bentonite is a natural material modified into inorganic bentonite (Zenith/Fe) by inserting Fe in the interlayer space of bentonite. The modified structures used in adsorption process compared to unmodified bentonite showed higher adsorption capacities of Zenith/Fe near Phoslock. Based on favorable adsorption, values of Zenith-N, Zenith/Fe, and Phoslock 1/n 0.965, 0.807, and 0.837 less than 1.0, respectively, followed the pseudo-second-order rate. As the most important contaminant, nitrogen is dissolved in the form of nitrate and formed inorganic pollutants that are highly stabile to make highly dangerous aerobic systems in the underground water. Nitrate contaminant has been found up to 223 mg/L in artesian wells, whereas the World Health Organization suggested that the maximum limit is 45 mg/L. Many studies reveal that high concentrations of this pollutant produce newborn diseases as an "infant cyanosis" (blue baby syndrome) and eliminated by adsorption techniques. Duran et al. employed natural calcium bentonites and it was modified by HCl and H_2SO_4 (thermoactivation) for nitrate removal in aqueous solutions. FTIR and Lambert–Beer law confirmed the concentration of nitrate in the solution, XRD characterized the size of clay particles and surface area measured by BET techniques. Ammonia is the major inorganic contaminate, and municipal, agricultural, and industrial wastes are the main sources that caused water pollution. Removal of ammonia is demandable due to decreased oxygen level in ponds, which are dangerous for river, fish, and other microbes. Therefore, modified bentonite is used for the removal of ammonium ion in water (Table 4.3) [101–109].

Table 4.3: Removal of inorganic impurities by bentonite clay-modified materials.

S. no.	Inorganic impurities	Bentonite clay and modified materials	References
1	Fluoride	Bentonite/chitosan beads Ca-bentonite and acid-treated bentonite Fe(III)-modified bentonite clay	[98–100]
2	Phosphorus	Al-modified bentonite clay	[101]
3	Phosphate	Modified bentonite	[102–103]
4	Nitrate	Zero-valent iron and pillared bentonite calcium montmorillonite activated by hydrochloric acid	[104]
5	Ammonium ion	Natural zeolite and bentonite, zeolitization of a bentonite, natural Turkish clinoptilolite	[105]
6	Ammonia-nitrogen	Al-Bent, Fe-Bent, CTMAB-Bent	[106–109]

4.5.4.3 Organic impurities

A numerous organic contaminates are found in water such as insecticides or pesticides; phenols, chlorophenols (CP) [110], and formaldehyde are organic, toxic, allergenic, and carcinogenic compounds show hazardous effects on environmental organisms. In general, organic contaminates (Table 4.4) are counted as major pollutants discharged from both industrial and nonindustrials. Salman et al. [111] studied the removal of formaldehyde using bentonite and kaolin clay materials and analyzed the uptake capacity of formaldehyde 5.03 mg/ g, kaolin and 3.41 mg/ g, bentonite as adsorbent with correlation coefficient linear Langmuir isotherm 0.995 and 0.985, respectively, which indicate that bentonite consumes greater uptake capacity than at pH 2–3. The considerable formaldehyde adsorption capacity and affinity from Freundlich isotherm of kaolin and bentonite (K_f = 0.363 and 0.468 mg/g) and adsorption intensity (n = 1.435 and 1.307), respectively. But the BT values for kaolin and bentonite, 0.062 and 0.088 kJ/mol, represented physisorption due to weak interaction between formaldehyde and adsorbents. Gu et al. [112] reported combined ozonation and bentonite coagulation process (COBC) for removal of humic acid (HA) and *o*-dichlorobenzene (DCB) from drinking water. When compared with

Table 4.4: Removal of organic impurities by bentonite clay-modified materials.

S. no.	Organic impurities	Bentonite clay and modified materials	References
1	Phenol	TMA-, BTMA-, TEA-, BTEA-, TMH-bentonites	[110]
2	Formaldehyde	Kaolin and bentonite	[111]
3	Humid acid and *O*-dichlorobenzene	Ozonation and bentonite coagulation	[112]
4	4-Chlorophenol and dichloroacetic acid	Ti-, Zr-, and Ti/Zr-pillared bentonites	[114]
5	Atrazine	4-Vinylpyridine-co-styrene montmorillonite	[115]
6	Chlorobenzene	CTMAB-modified bentonite and kaolinite	[121]
7	Salicylic acid, acetylsalicylic acid, and atenolol	Clinoptilolite, bentonite, and kaolin	[122]
8	Naproxen, salicylic acid, clofibric acid, and carbamazepine	Inorganic–organic intercalated bentonite modified with transition metals	[123]
9	Naphthalene and phosphate	CTMAB-Al13 intercalated bentonites	[123]

ozonation and coagulation, COBCs are highly effective in the removal process. To study this, HA and DCB are removed by catalytic ozonation and bentonite coagulation; HA removal is needed for coagulation method; DCB to oxidation process in COBC is highly effective in destroying aromatic chemicals. COBC bentonite improved the potentially advantageous dissolved pollutants in drinking water. CP and dichloroacetic (DCAA) are rich industrial toxic components and are recognized from different polluting sources like herbicides, pesticides, chemical and solvent manufacturing, and the paint industry. The existence of recalcitrant pollutants in aqueous atmosphere are crucial to avoid water recycle characterized by toxicity and low biodegradability [113], while DCAA is considered as animal carcinogens which was found in drinking water at 50 µg/L [114]. Atrazine (2-chloro-4-ethylamino-6-isopropylamino-striazine) is the chemical name of herbicides found in the USA [115]. The high concentrations are detected on ground surface of waters in Europe and North America [116–118] due to the ability to persist in soils, low sediment partitioning, slow rate of degradation. In the USA, the high limit of atrazine in drinking water is 3 PPB, whereas the European Union legislation banned to fix a limit of 0.1 ppb [119]. The removal of atrazine from contaminated water by granular-activated carbon (GAC) [120] proposed polymer–clay composites adsorbed on montmorillonite. Batch experiments demonstrated that the most appropriate composite poly(4-vinylpyridine-co-styrene)-montmorillonite, PVP-co-S90%-mont) removed 90–99% of atrazine (0.5–28 ppm). Column filter investigated 2 g of the PVP-co-S90%-montmorillonite with mixed composite and removed by sand 93–96% of atrazine (800 ppb). The dissolved organic matter (3.7 ppm) by GAC filter is to remove atrazine. The uncontrolled use of fertilizers such as chlorobenzene (CB) creates major problems on groundwater surface and produced unpleasant smell and taste at very low concentration. Chronic effects (long and short term) exposed central nervous system diseases including numbness, cyanosis, hyperesthesia (increased sensation) narcosis, restlessness, tremors, and muscle spasms in humans. Especially, fertilizer and its toxicity removed by clay minerals due to low cost of clay minerals with specific high surface areas is the main advantage for regular utility of natural adsorbent. Shu et al. [121] have examined sorption of CBs by cetyltrimethyl ammonium bromide (CTMAB)-modified bentonite and kaolinite. The linear CBs partitioning behavior loaded CTMAB level over the range of CB concentrations and linear sorption produced by surfactant-modified smectite and halloysite which showed 100% CEC. For sorption of CBs, CTMAB-kaolinite follow intradiffusion particles better than CTMAB-bentonite. As per thermal analysis, the values of Gibbs free energy ΔG^{o}, enthalpy change ΔH^{o}, and the entropy of the adsorption ΔS^{o} CBs sorption on CTMAB-bentonite and CTMAB-kaolinite was found in the range of −4.57 to −9.84 kJ/mol and −7.36 to −9.15 kJ/mol, 12.3–43.7 J/mol K, and 6.3–15.9 J/mol K, respectively. The results indicated that CBs strongly interacted with CTMAB-modified bentonite and kaolinite. Instead of that Rakic et al. [122] described the removal of salicylic acid, acetylsalicylic acid, and atenolol by water phase adsorption techniques onto natural zeolites and clays. Rivera-Jimanez

[123] modified inorganic–organic-intercalated bentonite (IOB) adsorbents for the removal of relevant emerging contaminants such as naproxen, salicylic acid, clofibric acid, and carbamazepine from water and materials characterized by XRD, porosimetry, SEM, TG analysis, and FTIR spectroscopy indicated general structural integrity. Consequently, the transition metal-modified IOBs displayed adsorption capacities and varied type of metal, pH, and nature of the adsorbents. The largest adsorption capacity was observed for salicylic acid, removed by modified hexadecyltrimethylammonium bromide. Organomodified clays are obtained from bentonite and kaolin involved ion-exchanged removed octadecyl dimethyl benzyl ammonium chloride, hydrophobic organic compounds, and oxyanions from polluted municipal wastewater. Mainly organobentonite used in wastewater treatment presents high concentration of hydrophobic organic compounds (HOCs) and landfill liner, whereas high affinity to oxyanions in water showed hydroxyl metal-pillared bentonites. Zhu et al. established bentonite sorbents and simultaneously removed HOCs and phosphate from water. CTMAB and hydroxy-aluminum (Al13) have been used to intercalate IOBs [107, 122, 123].

4.5.4.4 Pathogens

Pesticides are organic and carcinogenic compounds, evaluated through industrial, domestic, and agricultural wastes. For removal of pesticides, kaolinite is affected by adsorbents performed without the need for any organic modifications and showed similar adsorption behavior as malathion. The adsorption/desorption cycle of malathion regenerated and noticed kaolinite to decrease after the third cycles. Rauf et al. [100] reported agricultural insecticide endosulfan, which is very toxic for aquatic invertebrates, mammalian gonads, genotoxic and neurotoxic, and inserted bentonite materials for the removal of its toxicity. Morris et al. [124, 125] reported fine-grained particles of clay minerals, kaolinite and montmorillonite, used for the removal of microcystin-LR from waterbodies causing serious health hazard problems (Table 4.5).

Table 4.5: Removal of pathogenic impurities by bentonite clay-modified materials.

S. No	Pathogen	Bentonite clay and modified materials	References
1	Endosulfan	Kaolinite	[100]
2	Malathion pesticides	Kaolinite modified	[100]
3	Microcystin-LR	Natural clay minerals consisting of kaolin and montmorillonite	[124]

4.5.5 As catalyst

The advance materials of bentonite as worked heterogeneous photocatalyst and suitable for treating low concentrations of pollutant such as TiO_2 Degussa P-25 used photocatalyst and showed low photonic efficiency. Ti/Zr-pillared montmorillonites have been prepared from natural bentonite characterized by UV-Vis DRS and XRD. The photocatalytic activities established for the removal of 4-CP and DCAA in water under the influence of preparation conditions. The photocatalytic activities are increased by the added amount of zirconium in pillorying procedure and are calculated at 673 K or microwaves (MW). The results showed that the Ti/Zr-pillared clays are more efficiently calcined through MW slightly higher than TiO_2 Degussa P-25 and removed 90% impurity of 4-CP [126, 127].

4.5.6 Nanocomposite materials

The modified form of bentonite materials is significantly used as polymer nanocomposite hybrid materials comprising organic polymer matrix and dispersed inorganic nanofillers obtained at very low dimension levels (nanoscale). The sol–gel, intercalation, exfoliation–adsorption, melt intercalation, and template methods were applied for the formation of polymer silicate nanocomposites such as polymethylmethacrylate, polystyrene, styrene-co-acrylonitrile, epoxy- and chitin-based polyurethane bionanocomposites integrating bentonite clay. The advantages of composite materials are their large surface area, strength, heat resistance, decreased gas permeability and flammability, enhanced its property and described by optical microscopy gives reveled results of CEC [128–134].

4.5.7 In medical implements

Naturally present clay materials (calcium bentonite) are used for many purposes externally and internally such as skin care, clay facials, clay bathing, poultices and body wraps, cosmetic, internal diseases, intestinal and genital infections, nanodrugs (nanoclay) in pharmaceuticals. The powers of an aqueous solutions of bentonite and its related minerals control and prevent infectious diseases; for examples, hydrated aluminum silicate (diarrheas and cholera), hydrated bentonite detoxifies the human alimentary canal, and kaolin as a supplement to animals and prevent diarrheal diseases released from pigs. Instead, clay minerals are extensively used as excipient and active agents as drug–clay interaction is possible to the modified drug at nanoscale level and improves affinity of bioactive molecules by adsorptive features [135–138].

4.6 Summary and future direction

The various combinations of minerals in bentonite materials are used significantly in many purposes because these materials constantly interact with the world's organisms. The considerable ambitious properties of natural bentonite, clay, and montmorillonite materials characterized multidirectional and modified for the removal of involved contaminants existing in trace amounts in drinking water, especially biocompatible sorption process performed by its adsorption capacity that promises active uptake of metals, organic and inorganic contaminants, antimicrobes, and others bonded to natural or modified polymer/nanocomposite-coated materials and holds undesirable toxic materials that are located in the ecosystem. The constituents of raw bentonite materials and their rheological behavior are most important to describe compacted soil structures and their physical and chemical properties enhanced mechanical, thermal, dimensional, and barrier properties, hence, significantly applied at nanoscale level to control environmental pollutions.

References

[1] Hosterman, J W., and Patterson, S H. Bentonite and Fuller's earth resources of the United States, United States Government Printing Office, Washington D.C., USA, 1992.

[2] Melchior, S. In-situ studies on the performance of landfill caps (compacted soil liners, geomembranes, geosynthetic clay liners, capillary barriers). No. CONF-970208–PROC. 1997, Land Contamination Reclamation, 1997, 5(3), 209.

[3] Hussein, R A M., Mohamed, S A., Elemam, A E., and Alhassan, A A. Assessment of emission and performance of compression ignition engine with varying injection timing, Journal of Engineering Computer Science, 2014, 15(1), 34.

[4] Luckham, P F., and Rossi, S. The colloidal and rheological properties of bentonite suspensions, Advances in Colloid and Interface Science, 1999, 82, 43.

[5] Zhang, P., Guan, Q., and Li, Q. Mechanical Properties of Plastic Concrete Containing Bentonite, Research Journal of Applied Sciences and Engineering Technology, 2013, 5(4), 1317.

[6] Lim, S C., Gomes, C., and Kadir, M Z A A. Characterizing of Bentonite with Chemical, Physical and Electrical Perspectives for Improvement of Electrical Grounding Systems, International Journal of Electrochemical Science, 2013, 8, 11429.

[7] Sadik, C., Albizane, A., and El-Amrani, I E. Processing and characterization of alumina – mullite ceramics, Journal of Materials and Environmental Science, 2013, 4(6), 981.

[8] Sun, D., Sun, W., Yan, W., and Li, J. Hydro-mechanical behaviours of highly compacted sand-bentonite mixture, Journal of Rock Mechanics and Geotechnical Engineering, 2010, 2(1), 79.

[9] Christman, M., Benson, C H., and Edil, T B. Geophysical study of annular well seals. Groundwater, Groundwater Monitoring Remediation, 2002, 22(3), 104.

[10] Estornell, P., and Daniel, D E. Hydraulic conductivity of three geosynthetic clay liners, Journal of Geotechnical Engineering., 1992, 118(10), 1592.

[11] Evans, J C. Vertical cutoff walls., *Daniel*, D.E., (*Ed.*), Geotechnical practice for waste disposal, Chapman & Hall, London, Vol. 430, 1993.

[12] Kajita, L S. An improved contaminant resistant clay for environmental clay liner applications. Clays and Clay Minerals, 1997, 45(5), 609.

[13] Katsumi, T., Ishimori, H., Ogawa, A., Yoshikawa, K., Hanamoto, K., and Fukagawa, R. Hydraulic conductivity of nonprehydrated geosynthetic clay liners permeated with inorganic solutions and waste leachates, Soils Found, 2007, 47(1), 79.
[14] Smith, J A., Bartelt-Hunt, S L., and Burns, S E. Sorption and permeability of gasoline hydrocarbons in organobentonite porous media, Journal of hazardous materials, 2003, 96(1), 91.
[15] Vieiraa, M G A., Netoa, A F A., Gimenesb, M L., and Silva, M G C D. Removal of nickel on Bofe bentonite calcined clay in porous bed, Journal of Hazardous Materials, 2010, 176, 109.
[16] Ray, S S., and Okamoto, M. olymer/layered silicate nanocomposites: a review from preparation to processing, Progress in Polymer Science, 2003, 28, 1539.
[17] Nagendrarappa, G. Organic synthesis using clay and clay-supported catalysts, Applied Clay Science, 2011, 53, 106.
[18] Dobrowski, A. Adsorption-From Theory to Practice, Advances in Colloid and Interface Science, 2001, 93, 135.
[19] Donko, M., Micusik, M., Omastova, M., Bujdak, J., and Chorvat, D. Spectral characterisation of new organic fluorescent dyes with an alkoxysilane moiety and their utilisation for the labelling of layered silicates, Chemical Papers, 2013, 67, 18.
[20] Konta, J. Textural variation and composition of bentonite derived from basaltic ash, Clays and Clay Minerals, 1986, 34(3), 257.
[21] Obaje, S O. An overview on nanofluids and applications, International Journal of science and Technology, 2013, 3(2), 151.
[22] Onal, M. Physicochemical properties of bentonites: an overview, Communications Faculty of Sciences University of Ankara Series, 2006, 52(2), 1.
[23] Moore, D M., and Reynolds, R C J. X-ray diffraction and the identification and analysis of clay minerals, 2nd edition, Oxford University Press, New York, 1997.
[24] Uddin, F. Clays, nanoclays, and montmorillonite minerals, Metallurgical and Materials Transactions, 2008, 39, 2804.
[25] Murray, H H. Applied clay mineralogy: occurrences, processing and applications of koline, bentonites, palygorskitesepiolite and common clays developments in clay science, 1st edition, Elsevier, 2007.
[26] Van Olphen, H. An introduction to clay colloid chemistry, Wiley Interscience, New York, 1977.
[27] C.T. Johnston, E. Tomba Cz, in: J.B. Dixon, D.G. Schulze Eds. Soil mineralogy with environmental applications, Soil Sci. Soc. America, Madison, 2002, 37.
[28] Tomba´ Cz, E., and Szekeres, M. Colloidal behavior of aqueous montmorillonite suspensions: the specific role of pH in the presence of indifferent electrolytes, Applied Clay Science, 2004, 27, 75.
[29] Yener, N., Bicer, C., Onal, N., and Sarikaya, Y. Simultaneous determination of cation exchange capacity and surface area of acid activated bentonite powders by methylene blue sorption, Applied Surface Science, 2012, 258, 2534.
[30] Tamayo A., Kyziol-Komosinska J., Sánchez M., Calejas P., Rubio J., Barba M. Characterization and properties of treated smectites. Journal of the European Ceramic Society, 2012, 32, 2831.
[31] Clem, A G., and Doehler, E W. Industrial Applications of Bentonite, American Colloid Company, Skokie, Illinois.
[32] Pinnavaia, T J. Intercalated Clay Catalysts, Science, 1983, 220(4595), 365–371.
[33] Cadena, F., Rizvi, R., and Peters, R W. Feasibility studies for the removal of heavy metal from solution using tailored bentonite, hazardous and industrial wastes, in Proceedings of the 22nd Mid-Atlantic Industrial Waste Conference, 1990, 77–94, Drexel University.
[34] Wang, Q., Tang, A M., Cui, Y J., Barnichon, J D., and Ye, W M. Investigation of the hydro-mechanical behaviour of compacted bentonite/sand mixture based on the BExM model, Computers and Geotechnics, 2013, 54, 46.

[35] Delage, P., Howat, M D., and Cui, Y J. The relationship between suction and swelling properties in a heavily compacted unsaturated clay, Engineering Geology, 1998, 50(1), 31.

[36] Llorate, A., Villar, M V., Senchez, M., Gens, A., Pintado, X., and Alonso, E E. Geotechnics, Mechanical behaviour of heavily compacted bentonite under high suction changes, 2003, 53(1), 27.

[37] Romero, E., Villar, M V., and Llorate, A. Thermo-hydro-mechanical behaviour of two heavily over consolidated clays, Engineering Geology, 2005, 81(3), 255.

[38] Llorate, A., and Villar, M V. Advances on the knowledge of the thermo-hydro-mechanical behaviour of heavily compacted "FEBEX" bentonite, Physics and Chemistry of the Earth, 2007, 32(8–14), 701.

[39] Gens, A., and Alonso, E E. A framework for the behavior of unsaturated expensive clay, Canadian Geotechnical Journal, 1992, 29(6), 1013.

[40] Aguse, S S., Arifin, Y F., Tripathy, S., and Schanz, T. Acta Geotech. Swelling pressure–suction relationship of heavily compacted bentonite–sand mixtures. *Acta.Geotech*. 8, 155–165 (2013), Doi: 10.1007/s11440-012-0189-0.

[41] Alonso, E E., Vaunat, J., and Gens, A. Modelling the mechanical behaviour of expensive clay, Engineering Geology, 1999, 54(1), 173.

[42] Cui, Y J., Aissa, M Y., and Delage, P. A model for the volume change behaviour of heavily compacted swellings clay, Engineering Geology, 2002, 64(2–3), 233.

[43] Grim, R E. Clay mineralogy, McGraw-Hill Book company, INC, 1953.

[44] Polakowski, C., Sochan, A., Bieganowski, A., Ryzak, M., Foldenyi, R., and Toth, I. Influence of the sand particle shape on particle size distribution measured by laser diffraction method, J. International Agrophysics, 2014, 28, 195.

[45] Stul, M S., and Leemput, L V. Particle size distribution, cation exchange capacity and charge density of deferrated montmorillonites, Clay Minerals, 1982, 17, 209.

[46] Baik, M H., and Lee, S Y. Colloidal stability of bentonite clay considering surface charge properties as a function of pH and ionic strength, Journal of Industrial and Engineering Chemistry, 2010, 16, 837.

[47] Tombacz, E., and Szekeres, M. Colloidal behavior of aqueous montmorillonite suspensions: The specific role of pH in the presence of indifferent electrolytes, Applied Clay Science, 2004, 27, 75.

[48] Lagaly, G., and Ziesmer, S. The coagulation of montmorillonite dispersions, Advances in Colloid and Interface Science, 2003, 100–102, 105.

[49] Shin, Y., Lee, D., Lee, K., Ahn, K., and Kim, B. Surface properties of silica nanoparticles modified with polymers for polymer nanocomposite applications, Journal of Industrial and Engineering Chemistry, 2008, 14, 515.

[50] Joo, H H., Lee, H Y., Guan, Y S., and Kim, J C. Colloidal stability and in vitro permeation study of poly(e-caprolactone) nanocapsule containing hinokitiol, Journal of Industrial and Engineering Chemistry, 2008, 14, 608.

[51] Missana, T., and Adell, A. On the applicability of DLVO theory to the prediction of clay colloids stability, Journal of Colloid and Interface Science, 2000, 230, 150.

[52] Avena, M J., Cabrol, R., and Pauli, C P D. Study of Some Physicochemical Properties of Pillared Montmorillonites: Acid-Base Potentiometric Titrations and Electrophoretic Measurements, Clays and clay minerals, 1990, 38(356).

[53] Delgado, A., Caballero, F G L., and Bruque, J M. On the Zeta Potential and Surface Charge Density of Montmorillonite, Journal of Colloid and Interface Science, 1986, 113, 203.

[54] Duran, J D G., Tejeda, M M R., Arroyo, F J., and Caballerp, F G L. Rheological and electrokinetic properties of sodium montmorillonites suspension I. Rheological properties and interparticle energy of interaction, Journal of Colloid and Interface Science, 2000, 229, 107.

[55] Niriella, D., and Carnahan, R P. Comparison Study of Zeta Potential Values of Bentonite in Salt Solutions, Journal of Dispersion Science and Technology, 2006, 27, 123.
[56] Min, S W., Rong, T L., Ying, Z M. et al. The Relation between Mineral Composition and Swelling Character of Swelling Soil, Rock Soil Mechanics, 1994, 15((1)), 11.
[57] Fu, X Y., Ping, G Y., and Ze, Y Z. Tests on swelling deformation of expansive soils, J.hydra. engin., 1997, 9, 27.
[58] Chang, M L., Chen, Z X., and Ze, Y Z. High pressure jet grouted enclosed well-test and application to Xiaolangdi project, Dam Observation of Geotechnical Tests, 1999, 23(3), 36.
[59] Rayhani, M H T., Yanful, E K., and Fakher, A. Physical modeling of desiccation cracking in plastic soils, Engineering Geology, 2008, 97, 25.
[60] Yesiller, N., Miller, C J., Inci, G. et al. Desiccation and Cracking Behavior of Three Compacted Landfill Liner Soils, Engineering Geology, 2000, 57, 105.
[61] Tay, Y Y., Stewart, D I., and Cousens, T W. Shrinkage and desiccation cracking in bentonite–sand landfill liners, Engineering Geology, 2001, 60, 263.
[62] Basnett, C., and Bruner, R., Clay desiccation of a single-composite liner system, Geosynthetics 93 Conference, Vancouver, 1329 (1993).
[63] Hewitt, P J., and Philip, L K. Problems of clay desiccation in composite lining systems, Engineering Geology, 1999, 53, 107.
[64] Chen, Y., Zhou, S., Hu, R., and Zhou, C. Estimating effective thermal conductivity of unsaturated bentonites with consideration of coupled thermo-hydro-mechanical effects, International Journal of Heat and Mass Transfer, 2014, 72, 656.
[65] Chen, Y., Zhou, C., and Jing, L. Modeling coupled THM processes of geological porous media with multiphase flow: theory and validation against laboratory and field scale experiments, Computers and Geotechnics, 2009, 36(8), 1308.
[66] Rutqvist, J., Zheng, L., Chen, F., Liu, H H., and Birkholzer, J. Modeling of coupled thermo-hydro mechanical process with links to geochemistry associate to bentonite backfilled repository tunnels in clay formation, J. Rock Mechanics and Rock Engineering, 2013. http://dx.doi.org/10.1007/s00603-013-0375-x.
[67] Gong, Z., Liao, L., Lv, G., W, X. A simple method for physical purification of bentonite, Applied Clay Science, 2016, 119(2), 294–300.
[68] Bukit, N., Frida, E., and Harahap, M H. Preparation Natural Bentonite in Nano Particle Material as Filler Nanocomposite High Density Poliethylene (HDPE), Chemical Materials Research, 2013, 3(13), 2225.
[69] Robert, AS., Cliff, TJ. The surface properties of clay minerals. EMU Notes in Mineralogy, 2011, 11, 337–373.
[70] Lin, S H., and Juang, R S. Heavy metal removal from water by sorption using surfactant-modified montmorillonite, Journal of Hazardous Materials, 2002, 92(3), 315.
[71] Krishna, B S., Murty, D S R., and Prakash, B S J. Thermodynamics of chromium (VI) anionic species sorption onto surfactant-modified montmorillonite clay, Journal of Colloid and Interface Science, 2000, 229(1), 230.
[72] Bailey, S E., Olin, T J., Bricka, R M., and Adrian, D D. A review of potentially low-cost sorbents for heavy metals, Water research, 1999, 33(11), 2469.
[73] Srinivasan, R. Advances in Application of Natural Clay and Its Composites in Removal of Biological, Organic, and Inorganic Contaminants from Drinking Water, Advanced Materials Science and Engineering, Article ID 872531, 17, 2011. Doi: 10.1155/2011 /872531.
[74] Arthur, G., Celelyi, A N., and Robert, D W. Industrial applications of bentonite, 272–283 (2010), Tenth National conference on clays and clay Minerals, http://www.clays.org/journal/archive/volume%2010/10-1-272.pdf.
[75] Chisholm, F. Bentonite in industry, Mines Magazine, 1960, 30.

[76] Panwar, E P., and Dr. Ameta, N K. Stabilization of Dune Sand with Bentonite and Lime, Electronic Journal of Geotechnical Engineering, 2013, 18, 2667. http://www.ejge.com/2013/Ppr2013.248alr.pdf.
[77] Churchman, G J. Formation of complexes between bentonite and different cationic poly electrolytes and their use as sorbents for non-ionic and anionic pollutants, Applied Clay Science, 2002, 21(3–4), 177.
[78] Breen, C. The characterisation and use of polycation-exchanged bentonites, Applied Clay Science, 1999, 15(1–2), 187.
[79] Radian, A., and Mishael, Y G. Characterizing and Designing Polycation – Clay Nano composites As a Basis for Imazapyr Controlled Release Formulations, Environmental science and technology, 2008, 42(5), 1511.
[80] Zadaka, D., Nir, S., Radian, A., and Mishael, Y G. Atrazine removal from water by polycation – clay composites: effect of dissolved organic matter and comparison to activated carbon, Water research, 2009, 43(3), 677.
[81] Darder, M., Colilla, M., and Ruiz-Hitzky, E. Chitosan – clay nanocomposites: application as electrochemical sensors, Applied Clay Science, 2005, 28(1–4), 199.
[82] Darder, M., Blanco, M L., Aranda, P., Aznar, A J., Bravo, J., and Ruiz-Hitzky, E. Micro fibrous Chitosan – Sepiolite Nanocomposites, Chemistry of Materials, 2006, 18(6), 1602.
[83] Ruiz-Hitzky, E., Darder, M., and Aranda, P. Functional biopolymer nanocomposites based on layered solids, Journal of Materials Chemistry, 2005, 15(35–36), 3650.
[84] An, J H., and Dultz, S. Adsorption of tannic acid on chitosan-montmorillonite as a function of pH and surface charge properties, Applied Clay Science, 2007, 36(4), 256.
[85] Li, J M., Meng, X G., Hu, C W., and Du, Adsorption of phenol, p-chlorophenol and p-nitrophenol onto functional chitosan, J. Bioresource Technology, 2009, 100(3), 1168.
[86] Bulut, Y., Akcay, G., Elma, D., and Serhatlı, I E. Synthesis of clay-based superabsorbent composite and its sorption capability, Journal of Hazardous Materials, 2009, 171, 717.
[87] Rezaei, B., Ghiaci, M., and Sedaghat, M E. A selective modified bentonite–porphyrin carbon paste electrode for determination of Mn (II) by using anodic stripping voltammetry, Sensors and Actuators B: Chemical, 2008, 131, 439.
[88] Sultanbayeva, G S., Holze, R., Chernyakova, R M., and Zh, U. Removal of Fe^{2+}-, Cu^{2+}-, Al^{3+}- and Pb^{2+}- ions from phosphoric acid by sorption on carbonate-modified natural zeolite and its mixture with bentonite, Microporous and Mesoporous Materials, 2013, 170, 173.
[89] Oliveira, L C A., Rios, R V R A., Fabris, J D., Sapag, K., Garg, V K., and Lago, R M. Clay – iron oxide magnetic composites for the adsorption of contaminants in water, Applied Clay Science, 2003, 22, 169.
[90] Bhattacharyya, K G., and Gupta, S S. Pb (II) uptake by kaolinite and montmorillonite in aqueous medium: influence of acid activation of the clays, Colloids and Surfaces A, 2006, 277, 191.
[91] Kaya, A., and Oren, A H. Adsorption of zinc from aqueous solutions to bentonite, Journal of Hazardous Materials, 2005, 125, 183.
[92] Sen, T K., and Khoo, C. Adsorption characteristics of zinc (Zn^{2+}) from aqueous solution by natural bentonite and kaolin clay minerals: a comparative study, Comprehensive Water Energy Environmental Engineering, 2013, 2, 1.
[93] Ayari, F., Srasra, E., and Ayadi, M T. Removal of lead, zinc and nickel using sodium bentonite activated clay, Asian Journal of Chemistry, 2007, 19, 3325.
[94] Mishra, P C., and Patel, R K. Removal of lead and zinc ions from water by low cost adsorbents, Journal of Hazardous Materials, 2009, 168, 319.
[95] Aytas, S., Yurtlu, M., and Donat, R. Adsorption characteristic of U (VI) ion onto thermally activated bentonite, Journal of Hazardous Materials, 2009, 172(2–3), 667.

[96] Zhao, G., Zhang, H., Fan, Q. et al. Sorption of copper (II) onto super-adsorbent of bentonite – polyacrylamide composites, Journal of Hazardous Materials, 2010, 173(1–3), 661.
[97] Maramis, S V., Kumiawan, A., Ayucitra, A., Sunarso, J., and Ismadji, S. Removal of copper ions from aqueous solution by adsorption using LABORATORIES-modified bentonite (organo-bentonite), Frontiers of Chemical Science and Engineering, 2012, 6, 58.
[98] Triantafyllou, S., Christodoulou, E., and Syngouna, P N. Removal of Nickel and Cobalt from Aqueous Solutions by Na-Activated Bentonite, Clays and Clay Minerals, 1999, 47, 567.
[99] Zhang, Y., Wang, D., Liu, B., Gao, X., Xu, W., Liang, P., and Xu, Y. Adsorption of Fluoride from Aqueous Solution Using Low-Cost Bentonite/Chitosan Beads, American Journal of Analytical Chemistry, 2013, 4, 48.
[100] Rauf, N., Tahir, S S., Kang, J H., and Chang, Y S. Equilibrium, thermodynamics and kinetics studies for the removal of alpha and beta endosulfan by adsorption onto bentonite clay, Chemical Engineering Journal, 2012, 192, 369.
[101] Sergany, M E L., and Shanableh, A. Phosphorus Removal using Al-modified Bentonite Clay – Effect of Particle Size, Advances in Biomedical Engineering, 2012, 6, 323.
[102] Kui, T Y., Fa, T Z., Tao, W G., Min, L Z., and Wen, L D. Epoxy ether cleaving reactions catalyzed by supporting Lewis acidic ionic liquid, Chinese Journal of Chemical Engineering, 2006, 6, 197.
[103] Zamparas, M., Gianni, A., Stathi, P., Deligiannaki, Y., and Zacharias, I. Removal of phosphate from natural waters using innovative modified bentonites, Applied Clay Science, 2012, 62–63, 101.
[104] Li, J., Li, Y., and Meng, Q. Removal of nitrate by zero-valent iron and pillared bentonite, Journal of Hazardous Materials, 2010, 174, 188.
[105] Duran, C J M., Kou, M R S., Lopez, T. et al. Nitrate removal using natural clays modified by acid thermo activation, Applications of Surface Science, 2007, 253, 5762.
[106] Venglovsky, J., Pacajova, Z., Sasakova, N., Vucemilo, M., and Tofant, A. Adsorption properties of natural zeolite and bentonite in pig slurry from the microbiological point of view, Veterinarni Medicina Czech, 1999, 44, 339.
[107] Ruiz, R., Blanco, C., Pesquera, C., González, F., Benito, I., and Lopez, J L. Zeolitization of a bentonite and its application to the removal of ammonium ion from waste water, Applied Clay Science, 1997, 12, 73.
[108] Karadag, D., Koc, Y., Turan, M., and Armagan, B. Removal of ammonium ion from aqueous solution using natural Turkish clinoptilolite, Journal of Hazardous Materials, 2006, 136, 604.
[109] Yan, L., Xin, X., Xu, Y., Yu, H., and Du, B., Bioinform. Biomed. Engin. 3rd International Conference (2009).
[110] Shen, Y H. Removal of phenol from water by adsorption–Flocculation using organobentonite, Water research, 2002, 3, 1107.
[111] Salman, M., Athar, M., Shafique, U., Rahman, R., Ameer, S., Zulfiqar, S., and Azeem, M. Removal of formaldehyde from aqueous solution by adsorption on kaolin and bentonite: a comparative study, Turkish Journal of Engineering and Environmental Sciences, 2012, 36, 263.
[112] Gu, L., Zhang, X., Lei, L., and Liu, X. Concurrent removal of humic acid and o-dichloro-benzene in drinking water by combined ozonation and bentonite coagulation process, Water Science and Technology, 2009, 60, 3061.
[113] Calvo, L., Mohedano, A F., Casas, J A., Gilarranz, M A., and Rodrýguez, J J. Microwave-induced regeneration of activated carbons polluted with phenol. A comparison with conventional thermal regeneration, Carbon, 2004, 42, 1377.
[114] Liu, Y., and Mou, S. Determination of bromate and chlorinated haloacetic acids in bottled drinking water with chromatographic methods, Chemosphere, 2004, 55(1253).

[115] Kiely, T., Donaldson, D., and Grube, A. Pesticide industry sales and usage: 2000 and 2001 market estimates, U.S.Environmental Protection Agency, Washington, DC, USA, 2004.
[116] Cerejeira, M J., Viana, P., Batista, S. et al. Pesticides in Portuguese surface and ground waters, Water research, 2003, 37(1055).
[117] Papastergiou, A., and Mourkidou, E P. Occurrence and spatial and temporal distribution of pesticide residues in groundwater of major corn-growing areas of Greece (1996–1997), Environmental science and technology, 2001, 35, 63.
[118] Graziano, N., Mcguire, M J., Roberson, A., Adams, C., Jiang, H., and Blute, N. 2004 National Atrazine Occurrence Monitoring Program Using the Abraxis ELISA Method, Environmental Science and Technology, 2006, 40, 1163.
[119] Sass, J B., and Colangelo, A. European Union Bans Atrazine, While the United States Negotiates Continued Use, International Journal of Occupational and Environmental Health, 2006, 12, 260.
[120] Zadaka, D., Nir, S., Radian, A., and Mishael, Y G. Atrazine removal from water by polycation–clay composites: effect of dissolved organic matter and comparison to activated carbon, Water research, 2009, 43, 677.
[121] Shu, Y., Li, L., Zhang, Q., and Wu, H. kinetics and thermodynamic studies for sorption of chlorobenzenes on CTMAB modified bentonite and kaolinite, Journal of Hazardous Materials, 2010, 173, 47.
[122] Rakic, V., Rajic, N., Dakovic, A., and Auroux, A. The adsorption of salicylic acid, acetylsalicylic acid and atenolol from aqueous solutions onto natural zeolites and clays: Clinoptilolite, bentonite and kaolin, Microporous and Mesoporous Materials, 2013, 166, 185.
[123] Jimenez, S M R., Lehner, M M., Lafaurie, W A C., and Maldonado, A J H. Removal of Naproxen, Salicylic Acid, Clofibric Acid, and Carbamazepine by Water Phase Adsorption onto Inorganic – Organic-Intercalated Bentonites Modified with Transition Metal Cations, Environmental Engineering Science, 2011, 28.
[124] Donia, A M., Atia, A A., Hussien, R A., and Rashad, R T. Comparative study on the adsorp- tion of malathion pesticide by different adsorbents from aqueous solution, Desalination and Water Treatment, 2012, 47, 300.
[125] Morris, R J., Williams, D E., Luu, H A., Holmes, C F B., Andersen, R J., and Calvert, S E. The adsorption of microcystin-LR by natural clay particles, Toxicon, 2000, 38(2), 303.
[126] Houari, M., Saidi, M., Tabet, D., Pichat, P., and Khalaf, H. The Removal of 4-chlorophenol and Dichloroacetic Acid in Water Using Ti-, Zr- and Ti/Zr-Pillared Bentonites as Photo catalyst, "American Journal of Applied Sciences, 2005, 2, 1136.
[127] Meleroa, J A., Bautistaa, L F., Iglesiasb, J., Moralesa, G., and Vazqueza, R S. Production of biodiesel from waste cooking oil in a continuous packed bed reactor with an agglomerated Zr-SBA-15/bentonite catalyst, Applied Catalysis B: Environmental, 2014, 145, 197.
[128] Zubera, M., Ziaa, K M., Mahboobb, S., Hassan, M., and Bhattic, I A. synthesis of chitin – bentonite clay based polyurethane bio-nanocomposites, International Journal of Biological Macromolecules, 2010, 47, 196.
[129] Lee, D C., and Jang, L W. Preparation and characterization of PMMA–Clay hybrid composite by emulsion polymerization, Journal of Applied Polymer Science, 1996, 61, 1117.
[130] Noh, M W., and Lee, D L. Synthesis and characterization of PS-clay nanocomposite by emulsion polymerization, Polymer Bulletin, 1999, 42, 619.
[131] Noh, M H., Jang, L W., and Lee, D C. Intercalation of styrene – acrylonitrile copolymer in layered silicate by emulsion polymerization, Journal of Applied Polymer Science, 1999, 74, 179.
[132] Noh, M H., and Lee, D C. Comparison of characteristics of SAN – MMT nanocomposites prepared by emulsion and solution polymerization, Journal of Applied Polymer Science, 1999, 74, 2811.

[133] Kim, J., Ahmed, R., and Lee, S J. Synthesis and linear viscoelastic behavior of poly(amic acid) – organoclay hybrid, Journal of Applied Polymer Science, 2001, 80, 592.
[134] Lee, K M., and Han, C D. Effect of hydrogen bonding on rheology of polycarbonate/ organoclay nanocomposites, Polymer, 2003, 4573.
[135] Oh, Y J., Choi, G., Choy, Y B., Park, J W., Park, J H., Lee, H J., and Choy, J H. Aripiprazole Montmorillonite: A New Organic – Inorganic Nanohybrid Material for Biomedical Applications, Chemistry, 2013, 19(15), 4869.
[136] Trckova, M., Matlova, L., Dvoraska, L., and Pavlik, I. Bentonite, and zeolites as feed supplements for animals: health advantages and risks. A review, Veterinarni Medicina Czech, 2004, 49, 389.
[137] Damrau, F., New York City, Pioneers in Neurology, www.digestivehealth.net.au
[138] Suresh, R., Boekar, S N., Sawant, V A., Shende, V S., and Dimble, S K. Nanoclay drug delivery system, International Journal of Pharmaceutical Sciences and Nanotechnology, 2010, 3(2), 901.

Felycia Edi Soetaredjo, Suryadi Ismadji, Yi-Hsu Ju,
and Aning Ayucitra

5 Removal of ammonium from aquatic environment using bentonite and its modified forms

5.1 Introduction

The presence of ammonia in an aquatic environment often creates serious problem to the aquatic ecosystem due to eutrophication process. The existence of this substance in certain amount in water environment can promote the excessive growth of algae and as a consequence reduce the concentration of dissolved oxygen in water [1]. Accumulation of ammonia in aquatic system occurs through various pathways such as natural byproducts of fish metabolism, microbial metabolism, agricultural operations, food processing industries, pulp and paper factories, fertilizer industries, and municipal wastewater discharge.

In aquaculture industry, one of the most important parameters of water quality is the ammonia content. This substance has a very bad impact to aquatic biota, especially fish and crustacean. At low concentration, ammonia can cause stress to fish and also damage gills and other fish tissues. Long-time exposure of low concentration ammonia to fish causes poor growth, and fish will be more susceptible to bacterial infections [2]. Reduced reproductive capacity and reduced growth of the young are other possible ecological impacts of the presence of ammonia in the aquatic environment.

In an aqueous system, ammonia exists in two different forms simultaneously. The unionized ammonia (NH_3) is more harmful to aquatic microorganisms than the ionized ammonia (NH_4^+). Both forms of ammonia usually are expressed as total ammonia nitrogen or TAN. Temperature and pH have significant influence on the forms of ammonia [2]. At high pH and temperature, the formation of NH_3 is more favored than that of NH_4^+. Even the proportion of NH_3 and NH_4^+ fluctuates with pH and temperature; however, the TAN in water may remain constant [3].

The presence of ammonia in aquatic system also has a critical role in the nitrogen cycle. In the water environment, this substance is usually rapidly transformed into other nitrogenous forms through several processes such as fixation, assimilation, ammonification, nitrification, and denitrification [3]. Among these processes,

Felycia Edi Soetaredjo, Suryadi Ismadji, Department of Chemical Engineering, Widya Mandala Catholic University Surabaya, Surabaya, Indonesia
Yi-Hsu Ju, Aning Ayucitra, Department of Chemical Engineering, National Taiwan University of Science and Technology, Taipei, Taiwan

https://doi.org/10.1515/9783110630800-005

nitrification is the most important process in preventing the accumulation of high concentration of ammonia in water [3].

There are several methods or processes for removing ammonia from water or wastewater such as a biological process [4], adsorption [1, 5–10], stripping [11], struvite precipitation [12], ion exchange membrane [13], and the most common and economical method to remove ammonia from wastewater is a biological process [4, 5]. Since bacterial activities in a biological process are strongly affected by the environmental condition and the fluctuation of operating conditions, the efficiency of the process also fluctuates [14]. Among the available methods for the treatment of water or wastewater containing ammonia, adsorption possibly is still the promising method for this purpose since it is selective, cheap, has high removal efficiency, and produces the least amount of toxic sludge [15].

This chapter discusses various aspects of the adsorption of ammonia from a water environment using clay materials especially bentonite. Case study of the adsorption of ammonia from an aquaculture system using bentonite and its modified form is also given in this chapter.

5.2 Adsorption of ammonia and ammonium from water using zeolites

In the environment, nitrogen is one of the most abundant elements and can be found in various chemical forms such as ammonium (NH_4^+), ammonia (NH_3), nitrate (NO_3^-), nitrogen gas (N_2), nitric oxide (NO), and organic nitrogen. One of the very important processes to convert nitrogen into various forms of chemical is the nitrogen cycle. The nitrogen cycle consists of various processes such as fixation, ammonification, nitrification, and denitrification.

Ammonia is a compound required by most organisms for synthesizing protein. This compound is naturally produced by microbial metabolism, waste product of animals, and so on. Since ammonia has a critical role in the nitrogen cycle, its presence in the environment is very important. Nevertheless, due to rapid development of industrialization, ammonia is also released to the environment by industrial and human activities, and the latter cause negative environmental impacts on some aquatic ecosystems [3].

The adsorption of ammonia and ammonium from water environment or aqueous solution has been widely studied by various researchers. Some of the studies employed clay materials as adsorbents while others attempted to use unconventional adsorbents for this role. The studies on the adsorption of ammonia from aqueous solution using zeolite and its modified forms as adsorbents are summarized in Table 5.1.

Table 5.1: Ammonia and ammonium removal from water, wastewater, and aqueous solution using zeolite and its modified forms.

Type of zeolite	Maximum adsorption capacity (mmol/g)	Operation condition (*T*, °C; pH)	Reference
Mg-modified zeolite	n.a.	(23–25; 5–10)	[4]
Yemeni natural zeolite	0.62	(25, 35, 45; 2–10)	[7]
Natural clinoptilolite zeolite	0.89	(25; –)	[16]
Clinoptilolite treated with NaOH	1.15		
Mechanically modified clinoptilolite	1.39		
Rice husk ash-synthesized zeolite Y	2.35	(–; 4–10)	[17]
Mordenite	0.84		
Serbian clinoptilolite	1.81	(25; –)	[18]
Croatian clinoptilolite	1.35		
Turkish clinoptilolite	0.45	(25, 40, 55; –)	[19]
Zeolite 13X	0.30	(23; –)	[20]
Natural Chinese zeolite	0.39	(25; –) (–; 6.2)	[21]
Thermal-treated zeolite	0.59		
Microwave-treated zeolite	0.64		
New Zealand clinoptilolite	0.37	(–; <7.5)	[22]
New Zealand mordenite	0.53		
Na-mordenite	1.31	(5–60; –)	[23]
Natural Iranian zeolite	0.74	(25; –)	[24]
Natural zeolite	1.20 to 2.25	(22–26; –)	[25]
Natural mordenite	0.30 to 3.70		
Natural zeolite	2.76	(20–50; 2–10)	[26]
Mordenite	0.97	(25; 7)	[27]
Synthesized zeolite	1.35	(25; 8)	[28]
Modified Clinoptilolite	0.115	(25; 5, 7, 9)	[29]
Synthesized zeolite	1.32	(25; 8)	[30]
Natural chabazite	2.30	(15–35; 3–6)	[31]

Table 5.1 (continued)

Type of zeolite	Maximum adsorption capacity (mmol/g)	Operation condition (T, °C; pH)	Reference
Modified chabazite	2.66	(21, 30, 40, 50; 4–9)	[32]
Turkish (Yıldızeli) zeolite	0.54		

From Table 5.1, it is shown that different types of zeolite have different adsorption capacities. The adsorption capacity of the natural zeolites for exchanging ammonium is highly dependent upon the type of zeolite as well as the operating conditions as given in Table 5.1. Zeolite is a microporous aluminosilicate mineral, which is porous and can accommodate various cations (Na^+, Ca^{2+}, Mg^{2+}, etc.) in its structure. These cations can be exchanged with other cations in contact in the solution. Depending on its geographical location, the chemical and mineralogical composition of a natural zeolite varies greatly. The variation of chemical and mineralogical composition strongly affects the adsorption capacity of zeolites.

As mentioned previously, ammonia in aqueous solution exists in two different forms according to the following reactions:

$$NH_4{}^+ + OH^- \leftrightarrow NH_3 + H_2O \tag{5.1}$$

$$NH_3 + H_3O^+ \leftrightarrow NH_4{}^+ + H_2O \tag{5.2}$$

During the adsorption process using zeolite, the following equilibrium reaction mechanisms may take place [23]:

$$NH_4{}^+ + Na^+ - \text{zeolite} \leftrightarrow Na^+ + NH_4{}^+ - \text{zeolite} \tag{5.3}$$

$$H_2O + Na^+ - \text{zeolite} \leftrightarrow Na^+ + OH^- + H^+ - \text{zeolite} \tag{5.4}$$

$$H_2O + NH_4{}^+ - \text{zeolite} \leftrightarrow NH_4{}^+ + OH^- + H^+ - \text{zeolite} \tag{5.5}$$

$$NH_4{}^+ + H^+ - \text{zeolite} \leftrightarrow H^+ + NH_4{}^+ - \text{zeolite} \tag{5.6}$$

$$NH_3 + H^+ - \text{zeolite} \leftrightarrow NH_4{}^+ - \text{zeolite} \tag{5.7}$$

Depending on the acid sites of zeolite, NH_4^+ on the ion-exchange sites of zeolite possessing strong acid sites is less likely subject to hydrolysis than NH_4^+ on zeolites possessing weak acid sites [23].

Clinoptilolite zeolite ($(NaK_3)(Al_6Si_{40}O_{96}) \cdot 24H_2O$) is one of the naturally existing zeolites [16] and is probably the most widely studied for the removal of ammonia or ammonium from aqueous solution. Compared with other types of zeolite (mordenite, erionite, chabazite, phillipsite, analcime, laumonite, natrolite, and faujasite), clinoptilolite has the lowest theoretical cation exchange capacity (CEC). CEC of several types of common zeolites with their molecular structure [33] are given in Table 5.2. Even CEC of clinoptilolite is the lowest; however, it generally shows a high selectivity for NH^{4+} ion [16]. The selectivity of ion exchange on clinoptilolite as reported by Jha and Hayashi [16] was in the order of $K^+ > NH_4^+ > Na^+ > Ca^{2+} > Mg^{2+}$.

Table 5.2: Theoretical cation exchange capacity (CEC) of common zeolites [33].

Zeolite type	Molecular structure	Theoretical CEC (meq/g)
Clinoptilolite	$(NaK_3)(Al_6Si_{40}O_{96}) \cdot 24H_2O$	2.2
Mordenite	$(Na_8)(Al_8Si_{40}O_{96}) \cdot 24H_2O$	2.2
Erionite	$(Na_3Ca_3K_2)(Al_9Si_{27}O_{72}) \cdot 27H_2O$	2.8
Faujasite	$(Na_{58})(Al_{58}Si_{134}O_{384}) \cdot 240H_2O$	3.6
Chabazite	$(Na_6K_6)(Al_{12}Si_{24}O_{72}) \cdot 40H_2O$	3.9
Phillipsite	$(Na_5K_6)(Al_{10}Si_{22}O_{64}) \cdot 20H_2O$	4.5
Laumonite	$(Ca_4)(Al_8Si_{16}O_{48}) \cdot 16H_2O$	4.3
Analcime	$(Na_{16}(Al_{16}Si_{32})O_{96}) \cdot 16H_2O$	4.5
Natrolite	$(Na_{16})(Al_{16}Si_{24}O_{80}) \cdot 16H_2O$	5.3

The comparability of the adsorption capacity of mordenite and clinoptilolite has been reported by Weatherley and Miladinovic [22]. In their study, they also compared the adsorption performance of both natural zeolites for the adsorption of ammonia in the presence of other cations (Mg^{2+}, Ca^{2+}, and K^+). They found that the presence of these ions reduced the adsorption capacity of both natural zeolites toward ammonium. The effect is greatest for calcium ion and least for magnesium ion [22].

Compared to other natural zeolites, the adsorption performance of chabazite is less studied due to the availability in nature. Leyva-Ramos et al. [31] studied the removal of ammonium from aqueous solution using natural chabazite and its modified forms. In order to increase the adsorption capacity they used the hydrothermal treatment method to modify their sample with NaCl and KCl as modifying agents. The modification of chabazite with NaCl solution increased the adsorption capacity for exchanging ammonium; however, the modification with KCl solution gave the opposite result.

To design and analyze a proper ammonium adsorption system requires information about the adsorption equilibria [34]. In most cases, the adsorption experimental data of ammonia or ammonium onto zeolites were correlated using two-parameter isotherm models such as Langmuir and Freundlich. Both models were initially developed to represent gas phase adsorption equilibria and later has been adopted to correlate liquid phase adsorption experimental data.

Langmuir isotherm is one of the most widely used models to represent the adsorption data in liquid phase system. This isotherm is based on three assumptions, namely adsorption is limited to monolayer coverage, all surface sites are alike and each site can only accommodate one adsorbed atom, and the ability of a molecule to be adsorbed on a given site is independent of its neighboring site occupancy [34]. The Langmuir adsorption isotherm model has the form as follows:

$$q_e = q_m \frac{K_L C_e}{1 + K_L C_e} \tag{5.8}$$

where q_e and q_m are the amount adsorbed at equilibrium condition and the adsorption capacity of the adsorbent, respectively. Parameter K_L describes the adsorption affinity, while C_e is the concentration at equilibrium condition. In many cases, this model can describe the adsorption experimental data very well [1, 16, 17, 19, 24, 26, 27, 32]. One of the advantages of Langmuir model is that it is valid in a wide range of concentration since it has Henry law at very low concentration and also possesses the saturation limit capacity at high concentration.

Freundlich equation is an empirical equation, which was developed for adsorption in a heterogeneous system. Since this model is an empirical one, it can describe the adsorption experimental data in most systems. The mathematical form of Freundlich equation is

$$q_e = K_F {C_e}^{1/n} \tag{5.9}$$

where K_F is the Freundlich constant, n is the characteristic constant related to adsorption intensity or degree of favorability of adsorption. The parameter n also expresses the heterogeneity of the system. The more deviation the value of n from 1 the more heterogeneous the system is. A value of $1/n$ equal to 1 indicates linear adsorption leading to identical adsorption energies for all sites [34]. As a robust model, this equation is able to fit the experimental data of ammonium adsorption into different types of zeolites very well [25, 27, 28]. On several systems, this equation fails to conform to the adsorption experimental data due to the lack of Henry's law and saturation limit capacity [1, 16, 17, 19, 24, 26, 32].

5.3 Bentonite as promising adsorbent for ammonia adsorption

Bentonite is one of the clay minerals, which comprises largely of montmorillonite. The structure of bentonite is composed of microscopic platelets, which are stacked one on top of the other. These platelets consist of layers of aluminum hydroxide held between layers of silicate particles. Bentonite is generated frequently from the alteration of volcanic ash. Depending on the nature of their genesis, bentonite contains a variety of accessory minerals in addition to montmorillonite. These minerals may include quartz, feldspar, calcite, and gypsum. The presence of these minerals can affect the industrial value of a deposit, reducing or increasing its value depending on the application. Several types of bentonite are present in nature, and their classification depends on the dominant elements such as K, Na, Ca, and Al. Bentonite has special properties such as hydration, swelling, water absorption, and thixotropic that make this clay a valuable material for a wide range of uses and industrial applications. Bentonite is commonly used in foundry [35], composite material [36], wastewater treatment [37–40], drilling [41], oils/food [42–44], antibacterial [45], cosmetics [46], medicals [47, 48], catalyst [49, 50], and others.

The chemical nature and pore structure of bentonites generally determine their adsorption ability [51, 52]. Nevertheless, because of the hydrophilicity induced by the exchangeable metal cations, natural clays usually are not effective in adsorbing organic compounds. However, to improve the removal efficiency and adsorption capacity of these naturally occurring adsorbents, modification of surfaces has been insistently recommended and investigated [53–60]. There are two types of surface modification: Impregnation of organic molecules on bentonite surface which is characterized as a physical process, and organofunctionalization or grafting of organic molecules on bentonite surfaces which is characterized as a chemical process. Impregnation or organic modification process is accomplished through the replacement of inorganic exchangeable cations, such as Na^+, K^+, Al^{3+}, and Ca^{2+}, within the bentonite crystalline structure with organic cations, typically with quaternary ammonium cations. The surface of clays can be changed from hydrophilic to hydrophobic or organophilic by organofunctional molecules with surface hydroxyl groups, Lewis and Bronsted acidic sites, and so on by grafting organic groups on the clay surface [61]. The adsorption ability of bentonite can also be increased by strong inorganic acid treatment. The process is usually carried out at high temperature. When bentonites are acid-activated by hot mineral acid solutions, hydrogen ions attack the aluminosilicate layers via the interlayer region. This process alters the structure, chemical composition, and physical properties of the bentonite while increasing its adsorption capacity [43]. The acid-activated bentonite is widely used for bleaching vegetable oils.

The use of bentonite in industries for wastewater treatment applications today is strongly recommended due to their local availability, technological feasibility, technology applications, and cost effectiveness. Numerous studies have investigated the use of organobentonite as potential sorbents for organic contaminants in a wide variety of environmental applications [53–60, 62–64]. These results revealed that bentonite clays are promising materials for this purpose. Although bentonite has similar adsorption capacity to zeolites, however, the use of bentonite and its modified forms to treat water and wastewater containing ammonia or ammonium is still scarce, only a few studies employed bentonite for that purpose [65–71].

The modification of bentonite into the composite material for removal of phosphate and ammonium from eutrophic waters was conducted by Zamparas et al. [65]. The composite material (Bephos^{TM}) was prepared by embedding Fe and Cu ions and humic acid (HA) in the interlayer space of a natural bentonite. The comparison of the X-ray diffraction (XRD) pattern of the pristine bentonite and the composite (Bephos^{TM}) indicates that in Bephos™ the layered structure was clearly absent and Fe, Cu, HA, and clay lamellas were strongly interconnected. In the adsorption experiment, the maximum efficiency of removal of ammonia (47.5%) using pristine bentonite was achieved at pH 7 and with increasing pH to 9 the efficiency of removal of ammonia drops to 14.5%. The maximum ammonium adsorption efficiency of Bephos™ reaches 73% at pH 7 and remains quasistable up to pH 9 where ammonium uptake was 72% [65]. The adsorption capacity of Bephos™ for ammonia was 202.1 mg/g (11.23 mmol/g). This value is much higher than the adsorption capacity of natural zeolites and its modified form is listed in Table 5.1. Kinetic data showed that 70% removal of ammonia was achieved within 30 min.

A combination of adsorption and photocatalysis process for removal of ammonia from aging leachate was studied by Cai et al. [66]. In order to increase the adsorption capacity of bentonite, modification of bentonite using synthetic surfactant (CTMAB) was carried out using a simple procedure. The use of TiO_2 without the combination with CTA-bentonite gave low removal of ammonia due to the acidic condition and low intensity of UV light. By combining the adsorption process using CTA-bentonite and photocatalytic process using TiO_2 as catalyst, around 41% of ammonia removal could be achieved.

5.4 Adsorption of ammonia from aqueous solution using bentonite

Bentonite used in this study was obtained from Pacitan, East Java, Indonesia. This bentonite is mainly Ca–Mg-type bentonite and is generally suitable as raw material for adsorbent and bleaching earth. In East Java, bentonite reserves can be found in several areas such as Pacitan, Ponorogo, Blitar, and Trenggalek, with a total reserve of more than 500 million tons. Currently, the major use of this material is for the purification of crude palm oil and as drilling mud. In addition, this material also has potential environmental applications such as wastewater treatment. Although East Java has huge reserves of bentonite, one of the main problem and challenge in its application is in the chemical nature and characteristics of this material. The chemical nature and characteristics of East Java bentonite vary greatly, even in the same location. Without studying these chemical and surface characteristics, it is difficult to produce modified bentonite with constant quality and reliability.

CEC of the bentonite used in this study was determined using ASTM C837-99 and the CEC value was 0.69/g. The CEC value of this bentonite is much lower than the theoretical value of CEC of natural zeolites given in Table 5.2. The point zero charge (pH_{pzc}) of the bentonite was determined by titration method [72], and the value of pH_{pzc} of the bentonite is 3.4. The result of elemental analysis of the bentonite is given in Table 5.3.

Table 5.3: Elemental analysis of bentonite from Pacitan, East Java, Indonesia.

Compound	Composition (%)
Al_2O_3	33.20
SiO_2	57.11
FeO	3.37
CaO	3.95
MgO	0.42
K_2O	0.21
Na_2O	1.52
MnO	0.22

The XRD pattern of the bentonite was measured using a Bruker DS Advance powder diffractometer at 40 kV, 40 mA, and a step size of 0.01° using CuKα as the source of radiation. The XRD pattern of the bentonite is shown in Figure 5.1.

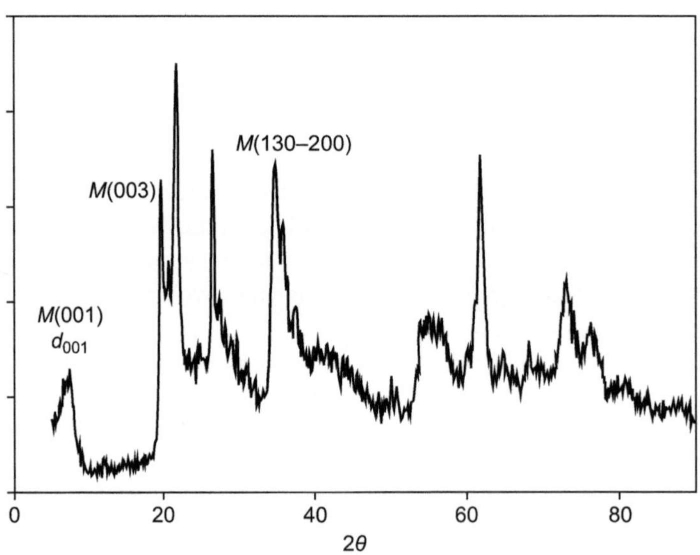

Figure 5.1: XRD pattern of bentonite from Pacitan, East Java, Indonesia.

The presence of montmorillonite in the bentonite was observed from reflections relative to the plane (001), (003), and (130–200) [73]. The interlayer spacing (d_{001}) of the bentonite was calculated using the Bragg equation as follows:

$$d = \frac{\lambda}{2\sin\theta} \tag{5.10}$$

where λ is the X-ray wavelength and θ is the scattering angle for the peak position. Since the radiation source was Ni-filtered CuKα the value of λ is 0.15405 nm. The diffraction peak (d_{100}) of the bentonite was observed at 2θ around 7.22° and this value corresponds to an interlayer spacing of 1.28 mm.

The pore structure of the bentonite was analyzed using nitrogen sorption analysis obtained at the boiling point of nitrogen gas (−196 °C). The nitrogen sorption analysis of the bentonite obtained at a relative pressure of 0.003–0.996 is shown in Figure 5.2.

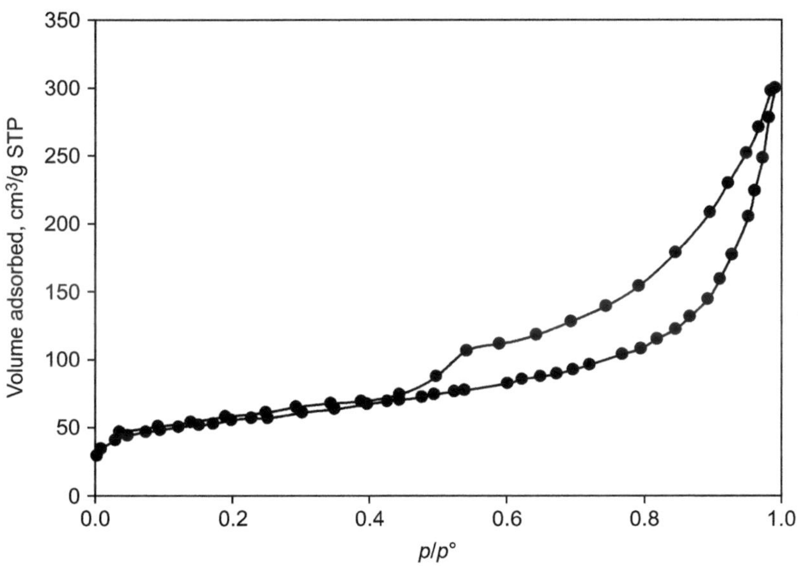

Figure 5.2: Nitrogen sorption analysis of bentonite from Pacitan, East Java, Indonesia.

The BET surface area of the bentonite was calculated using the standard BET equation at a relative pressure of 0.05–0.3. The BET surface area and pore volume of the bentonite are 186 m^2/g and 0.31 cm^3/g, respectively. Figure 5.2 clearly shows that the bentonite used possesses mesostructure. To confirm the structure of the bentonite, the density functional theory (DFT) analysis was conducted to determine the pore size distribution. The DFT analysis was conducted using medium regularization, and the result of the pore size distribution of the bentonite is shown in Figure 5.3. The

DFT analysis result confirms that bentonite from Pacitan has a mesoporous structure with pore size distribution mostly in the mesopores region (>20 A) with small amount of micropores as shown in Figure 5.3.

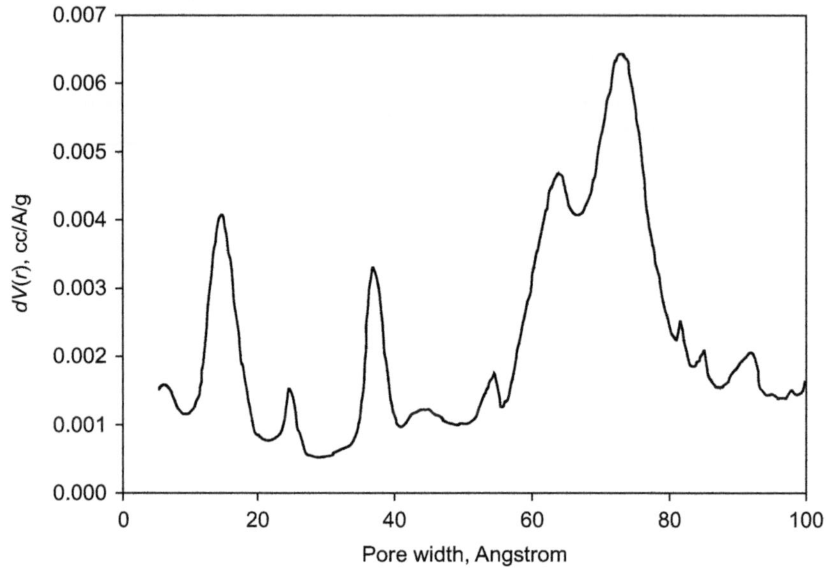

Figure 5.3: DFT pore size distribution of bentonite from Pacitan, East Java, Indonesia.

The surface functional groups in bentonite often play an important role in the adsorption process. The surface functional groups of the bentonite were characterized using the FTIR method and the result is summarized in Table 5.4.

Table 5.4: FTIR spectra of bentonite from Pacitan, East Java, Indonesia.

Functional groups	Wavenumber (1/cm)
O–H stretch for H_2O in the silica matrix	3,632
O–H stretch of silanol (Si–OH) groups	3,278
O–H bend, for adsorbed H2O at bentonite interlayer	1,648
Si–O–Si stretch of the tetrahedral sheet	1,029
Al–Al–OH bend	922
Al–O–Si bend (for octahedral Al)	698
Si–O–Si bend	436

The adsorption of ammonia from aqueous solution using bentonite as the adsorbent was conducted at three different temperatures (30, 40, and 50 °C). The temperature-dependent forms of adsorption isotherms were employed to correlate the adsorption experimental data. Brief descriptions about the temperature-dependent forms of several adsorption isotherm models [74] are described in subsequent paragraphs.

Langmuir model (eq. (5.8)) was developed based on the kinetic principle; the Langmuir parameters q_m and K_L have the temperature-dependent forms [74] as follows:

$$q_m = q_m{}^{\circ} \exp(\delta(T_o - T)) \tag{5.11}$$

$$K_L = K_L^{o} \exp\left[\frac{E}{RT}\right] \tag{5.12}$$

The maximum adsorption capacity at reference temperature T_o (K) is indicated by symbol $q_m{}^{\circ}$ (mg/g). The parameter δ in eq. (5.11) represents the coefficient expansion of the adsorbate and is usually in the order of 10^{-3}. $K_L{}^{\circ}$ (L/mg) and E (kJ/mol K) are the parameters representing the adsorption affinity and heat of adsorption at reference temperature, respectively.

The parameters of the Freundlich model (eq. (5.9)) have the mathematical forms [74] as follows:

$$K_F = K_F{}^{\circ} \exp\left[\frac{-\alpha RT}{A_o}\right] \tag{5.13}$$

$$\frac{1}{n} = \frac{RT}{A_o} \tag{5.14}$$

The adsorption capacity of the Freundlich model at reference temperature is indicated by parameter $K_F{}^{\circ}$ (mg/g)(mg/L)$^{-n}$. The adsorption potential and Clapeyron constants are represented by parameters α and A_o, respectively.

Sips equation is one of the widely used models to represent liquid phase adsorption equilibrium data. This model is also known as Langmuir–Freundlich equation, and has the form [74] as follows:

$$q_e = q_m\left[\frac{(K_s C_e)^{1/n}}{1 + (K_s C_e)^{1/n}}\right] \tag{5.15}$$

The temperature-dependent form of q_m follows eq. (5.11), while K_s (mg/L)$^{1/n}$ is the adsorption affinity constant. The parameter n in Sips model, similar to parameter in Freundlich equation, represents the heterogeneity of the system. In the case of n = 1, eq. (5.15) reduces to eq. (5.8). The temperature-dependent forms of eq. (5.8) are

$$K_S = K_S^o \exp\left[\frac{E}{RT_o}\left(1 - \frac{T}{T_o}\right)\right] \tag{5.16}$$

$$n = \frac{1}{\frac{1}{n_o} + \eta\left(1 - \frac{T_o}{T}\right)} \tag{5.17}$$

$K_s°$ and n_o are the adsorption affinity of the Sips equation at the reference temperature and heterogeneity of the system, respectively, while η is a constant which has no physical meaning.

Another equation that is suitable for submonolayer coverage system is the Toth equation. This model is also widely used for liquid phase adsorption. The Toth equation has the mathematical expression as follows:

$$q_e = q_m \frac{C_e}{\left(K_T C_e^t\right)^{1/t}} \tag{5.18}$$

The temperature-dependent forms of the Toth parameters are

$$K_T = K_T^o \exp\left[\frac{E}{RT_o}\left(\frac{T_o}{T} - 1\right)\right] \tag{5.19}$$

$$t = t_o + \gamma\left[1 - \frac{T_o}{T}\right] \tag{5.20}$$

$K_T°$ is the adsorption affinity at the reference temperature and t_o has the same meaning as that of n_o in Sips model.

The adsorption experiments were conducted in static mode at three temperatures. The initial concentration of the aqueous ammonium solution was 500 mg/L. The adsorption experiments were carried out at pH 4–10. The initial and equilibrium concentrations of ammonium in the solution were measured using a UV-Vis spectrophotometer (Shimadzu) [75]. In the adsorption of ammonium, pH is one of the most important parameters controlling the adsorption process at solid–liquid interface [65]. The effect of pH on ammonium removal from aqueous solution at 30 °C is depicted in Figure 5.4. The removal efficiency of ammonium from aqueous solution increased with the increase of pH from 4 to 7, and then decreased with the increase of pH as shown in Figure 5.4. At low pH, the ammonia is converted into the ammonium ions (eq. (5.2)), resulting in a substantial amount of ammonium ions available in the solution. Above pH 7, the ammonium ions are rapidly converted into ammonia (eq. (5.1)) and this condition becomes less favorable for the adsorption process since only ionized form of ammoniacal nitrogen can be removed by the adsorption process [76].

Since the maximum removal efficiency was achieved at pH 7, all adsorption experiments were conducted at this pH. The adsorption isotherms of ammonium onto bentonite and the fits of different adsorption models are given in Figures 5.5–5.8.

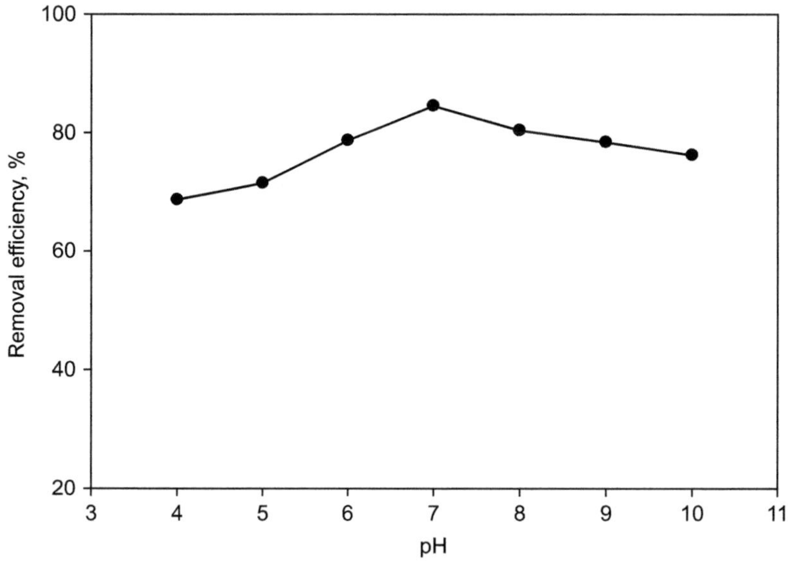

Figure 5.4: Effect of pH on adsorption of ammonium from aqueous solution using bentonite from Pacitan, East Java, Indonesia.

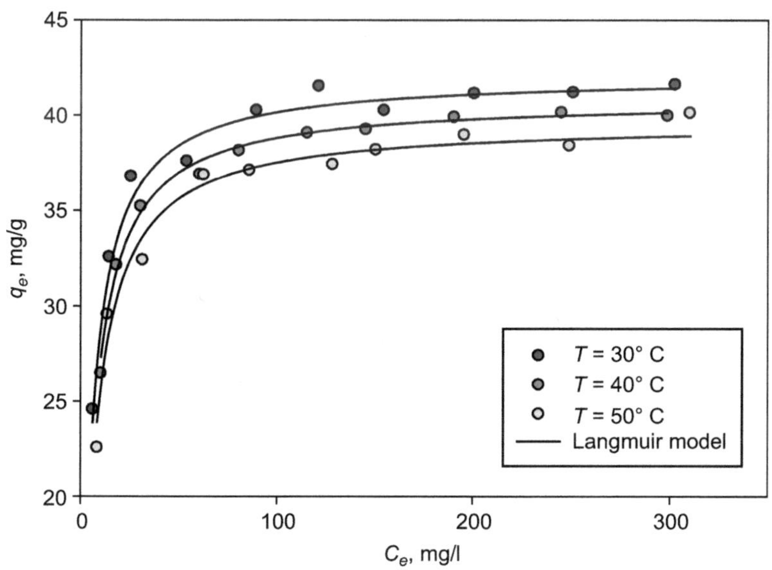

Figure 5.5: Adsorption isotherms of ammonium onto bentonite and fits of Langmuir equation.

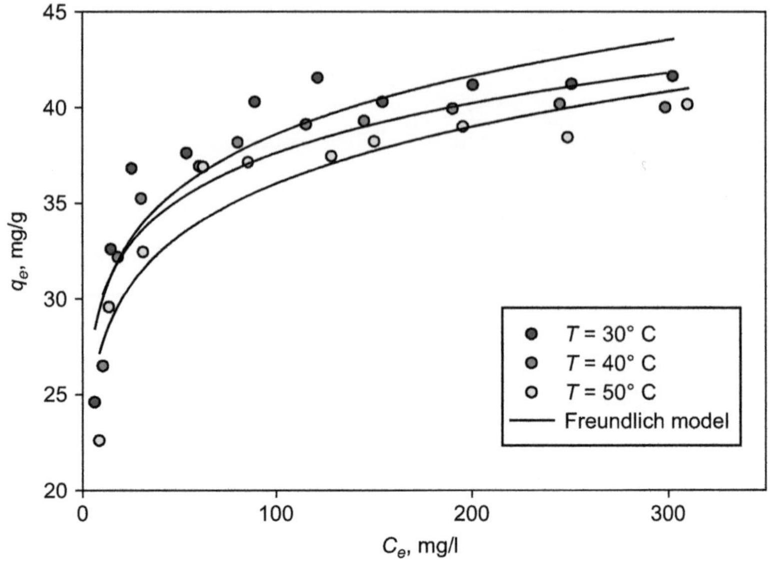

Figure 5.6: Adsorption isotherms of ammonium onto bentonite and fits of Freundlich equation.

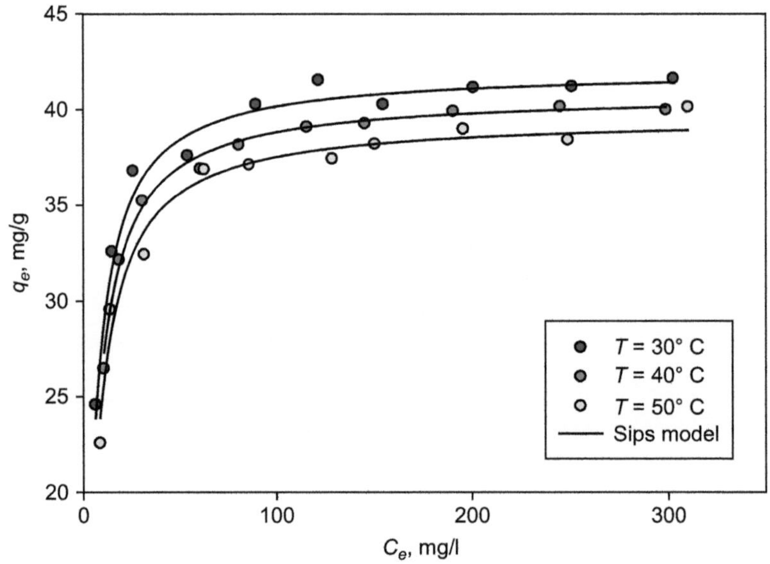

Figure 5.7: Adsorption isotherms of ammonium onto bentonite and fits of Sips equation.

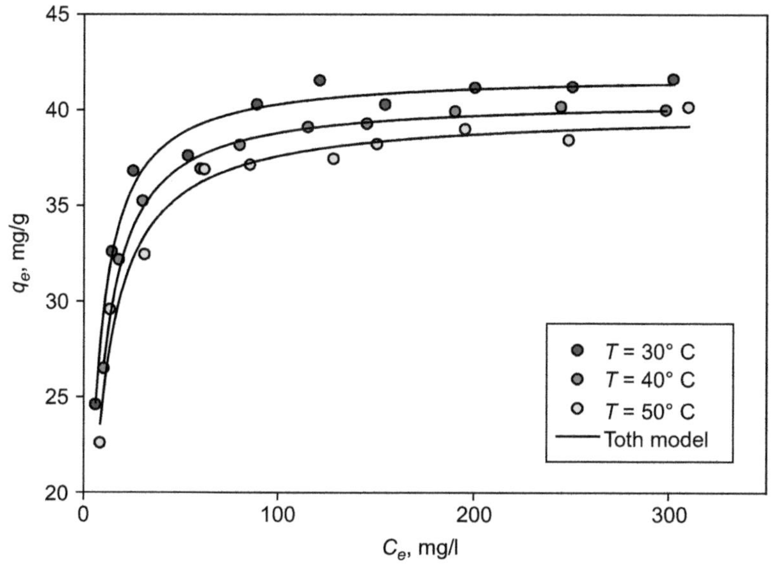

Figure 5.8: Adsorption isotherms of ammonium onto bentonite and fits of Toth equation.

A nonlinear least squares method was used to obtain the parameters of the adsorption equations. The fitted parameters of the models obtained from the nonlinear method are summarized in Table 5.5. The following objective function was minimized to obtain the best fitting of the equation parameters:

$$\mathrm{SSE} = \left[\frac{\left(\sum q_{e(\exp)} - q_{e(\mathrm{cal})}\right)^2}{n}\right]^{1/2} \tag{5.21}$$

Figure 5.6 clearly shows that the Freundlich model fails to represent the adsorption of ammonium onto bentonite. From Figures 5.5, 5.7, and 5.8, it can be seen that the Langmuir, Sips, and Toth describe the experimental data fairly well. Since Freundlich model fails to describe the experimental data, this model is not included in the subsequent discussion.

In order to determine the suitability of models to represent the adsorption experimental data, it is crucial to examine the physical meaning of each parameter [62] tabulated in Table 5.5. The adsorption capacity of the adsorbent at a reference temperature is represented by $q_m{}^o$. The values of this parameter for all equations (Langmuir, Sips, and Toth) are essentially the same as indicated in Table 5.5. The values of this parameter are reasonable and comparable to the values reported in literatures [17, 25, 26].

Further verification of the adequacy of the above isotherms is provided by comparing the value of parameter δ with the available values in the literatures. As

Table 5.5: The fitted temperature-dependent parameters of several isotherm models for adsorption of ammonia onto bentonite.

Isotherm	Parameters	Value
Langmuir	$q_m{}^o$, mg/g	42.7
	δ, 1/K	0.003
	$K_L{}^o$, L/mg	0.0142
	E, kJ/mol	6.810
	R^2	0.998
Freundlich	$K_F{}^o$ (mg/g)(mg/L)$^{-n}$	22.140
	α	0.154
	A_o	10.388
	R^2	0.871
Sips	$q_m{}^o$, mg/g	41.22
	δ, 1/K	0.003
	$K_s{}^o$, (mg/L)$^{1/n}$	0.0128
	E, kJ/mol	6.09
	n_o	1.0136
	H	0.0029
	R^2	0.995
Toth	$q_m{}^o$, mg/g	40.79
	δ, 1/K	0.003
	$K_T{}^o$	0.0111
	E, kJ/mol	6.29
	n_o	1.0327
	Γ	0.0029
	R^2	0.993

mentioned earlier, this parameter represents the expansion coefficient of adsorbate. The δ values of the ammonium ion are consistent with the values for many liquids [77, 78] although the precise value for the ammonium used in this study is not readily available.

The strength of the adsorbate molecule attached to the surface is measured by the affinity parameter (K_L° for Langmuir equation, K_s° for Sips model, and K_T° for Toth equation). The high value of affinity parameter indicates that the adsorbate molecules are strongly attached to the surface of adsorbent and the surface is covered by a greater number of adsorbate molecules [74]. From Table 5.5, it can be seen that all isotherm equations (Langmuir, Sips, and Toth) gave a correct and reasonable value of the affinity coefficient. A reasonable correlation between adsorption capacity parameter and affinity parameter is observed from Table 5.5. The highest fitted value of the affinity parameter was obtained from the fitting using Langmuir model and highest fitted value of adsorption capacity was also observed in Langmuir model. The lowest values of fitted affinity parameter and adsorption capacity were obtained from the fitting of the data using Toth model. Although the affinity parameter and adsorption capacity by Langmuir, Sips, and Toth were slightly different; however, the values of these parameters are still consistent with their physical meaning.

The next parameter is E (heat of adsorption). The value of E determines the type of adsorption. Physical adsorption processes usually have adsorption energies less than 40 kJ/mol, while higher energies (40–800 kJ/mol) suggest the involvement of chemisorption [62]. In physical adsorption, the bonding between adsorbate and adsorbent occurs through van der Walls force (physical bonds) while in the chemisorption the bonding between adsorbate and adsorbent is through chemical bonds. In physical adsorption, the uptake of adsorbate molecule decreases with increasing temperature, while chemisorption temperature has a positive effect on the uptake of adsorbate. The values of E obtained from different equations indicate that the adsorption of ammonium onto bentonite is the physical adsorption, and this phenomenon is supported by adsorption experimental data, and the uptake of ammonium ions decreased with increasing temperature.

From evaluating the physical meaning of the fitted parameters of Langmuir, Sips, and Toth equations, it is obvious that all equations can describe the adsorption experimental fairly well. It is not surprising because Sips and Toth are modified forms of Langmuir model.

5.5 Adsorption of ammonia from aqueous solution using modified bentonite

In order to increase the adsorption capability of bentonite, modification using sodium hydroxide was conducted under microwave irradiation. A brief description about the modification of bentonite using microwave technique is as follows: 50 g of pulverized bentonite were mixed with 250 mL of NaOH solution (1 M) and stirred at constant speed (500 RPM) for 6 h. The mixture was then irradiated for 10 min at

700 W. After the thermal irradiation was completed, the solid was separated from the mixture and repeatedly washed with distilled water until neutral pH. The modified bentonite was dried in an oven at 105 °C for 24 h.

The results of SEM (scanning electron microscopy) analysis of pristine bentonite and modified bentonite are given in Figure 5.9.

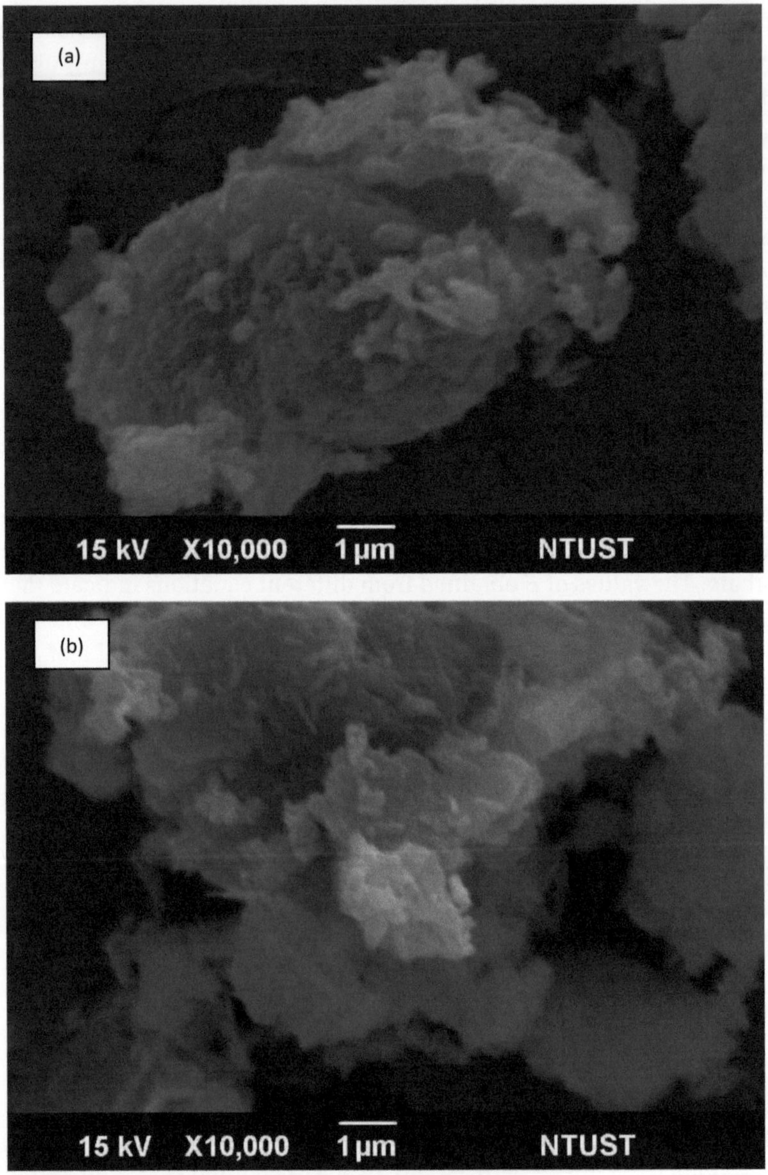

Figure 5.9: SEM analysis of (a) bentonite and (b) modified bentonite.

In general, modification using 6 M NaOH solution combined with microwave irradiation did not change the surface topography of the bentonite as indicated in Figure 5.9.

The XRD pattern of the modified bentonite is given in Figure 5.10. The presence of montmorillonite in the modified bentonite was still observed from reflections relative to plane (001), (003), and (130–200). The diffraction peak (d_{100}) of the modified bentonite was observed at 2θ around 5.63° and based on eq. (5.10) this diffraction peak corresponds to an interlayer spacing of 1.57 nm. The modification with NaOH increases the interlayer spacing from 1.28 to 1.57 nm. The increase in the basal/interlayer spacing of the modified bentonite indicates that sodium molecules were partially intercalated into the interlayer spaces in the bentonite structure, leading to an expansion in the interlamellar spacing of the bentonite.

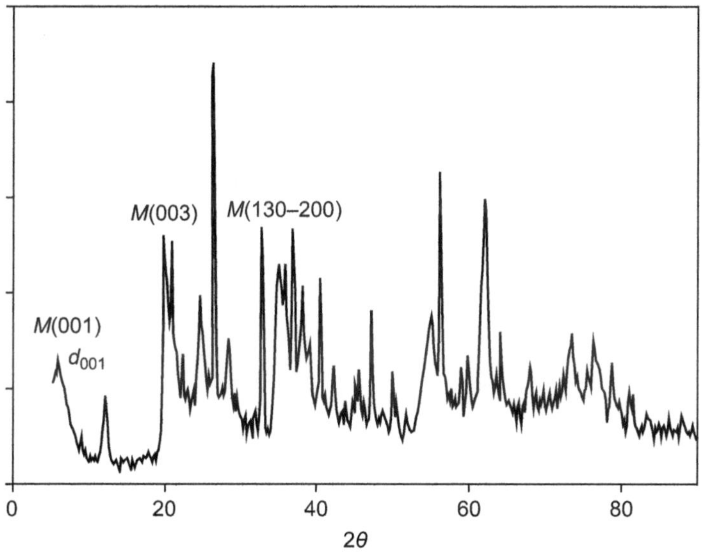

Figure 5.10: XRD pattern of NaOH-modified bentonite.

The nitrogen sorption isotherms of modified bentonite are depicted in Figure 5.11. The modified bentonite still exhibits mesoporous structure as indicated by the hysteresis between adsorption and desorption curves at relative pressures of 0.25–0.99. The BET surface area and pore volume of the modified bentonite are 165 m^2/g and 0.252 cm^3/g, respectively. The pore size distribution of modified bentonite was also determined by the DFT method, and the result is depicted in Figure 5.12. This figure clearly shows that the pore structure of the bentonite has changed during the modification with NaOH.

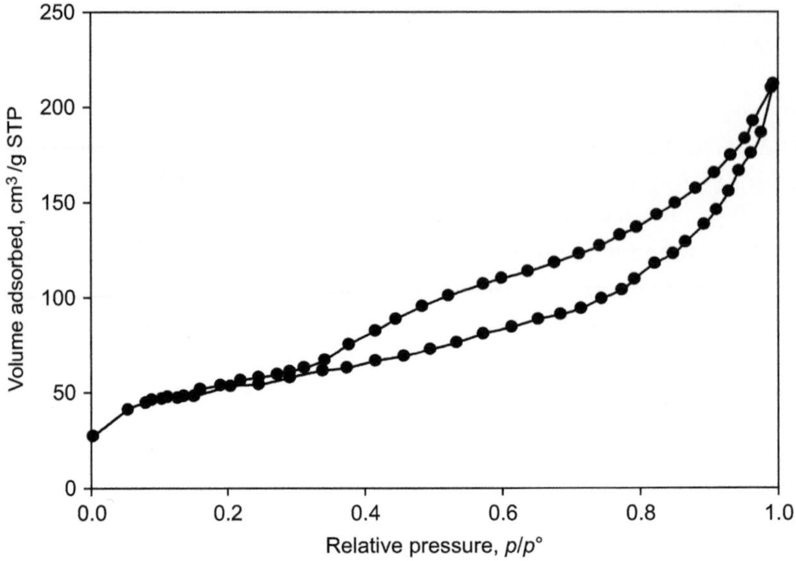

Figure 5.11: Nitrogen sorption analysis of modified bentonite.

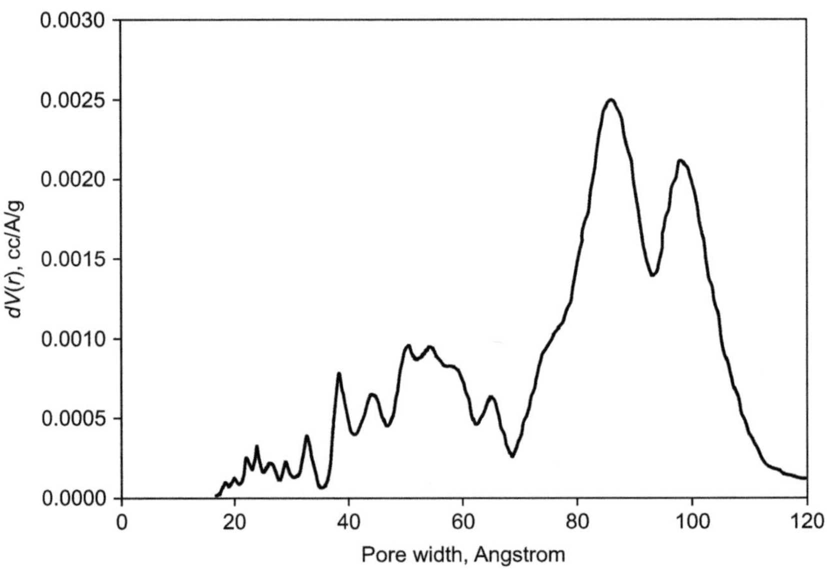

Figure 5.12: DFT pore size distribution of modified bentonite.

The adsorption experiment for removing ammonium from aqueous solution using NaOH-modified bentonite was also conducted in static mode with similar operating conditions of those experiments using bentonite as the adsorbent. The plots of adsorption experimental data and the isotherm models (Langmuir, Sips, and

Toth) are shown in Figures 5.13–5.15. The fitted temperature-dependent parameters of several isotherm models for the adsorption of ammonia onto NaOH-modified bentonite are given in Table 5.6.

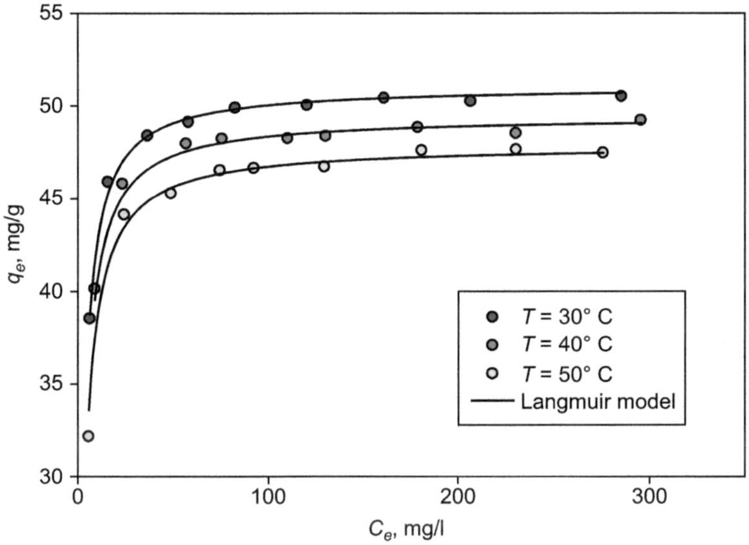

Figure 5.13: Adsorption isotherms of ammonium onto modified bentonite and fits of Langmuir equation.

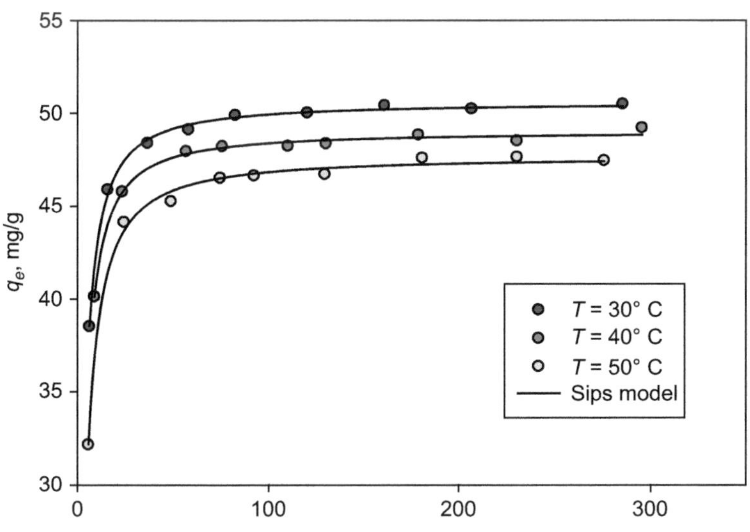

Figure 5.14: Adsorption isotherms of ammonium onto modified bentonite and fits of Sips equation.

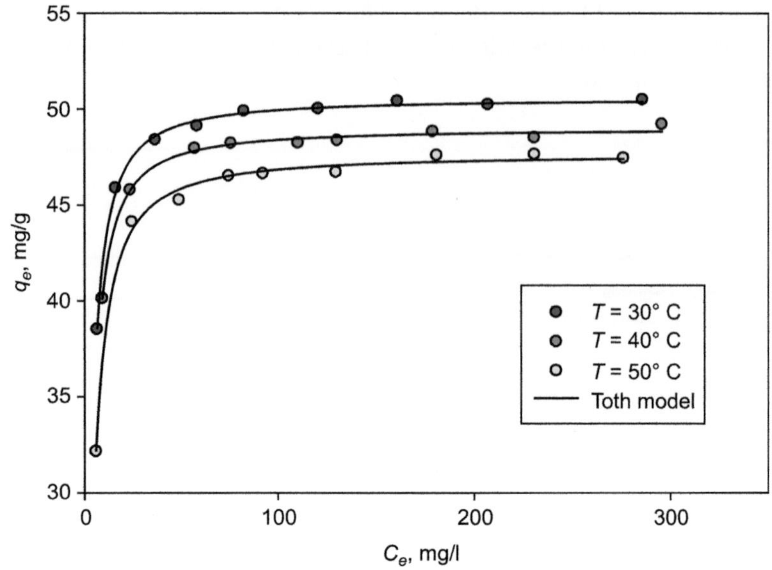

Figure 5.15: Adsorption isotherms of ammonium onto modified bentonite and fits of Toth equation.

Table 5.6: Fitted temperature-dependent parameters of several isotherm models for adsorption of ammonia onto modified bentonite.

Isotherm	Parameters	Value
Langmuir	q_m^o, mg/g	51.92
	δ, 1/K	0.0032
	K_L^o, L/mg	0.0183
	E, kJ/mol	8.35
	R^2	0.997
Sips	q_m^o, mg/g δ, 1/K	50.58 0.0029
	K_s^o, $(mg/L)^{1/n}$	0.0175
	E, kJ/mol	8.22
	n_o	1.068
	η	0.0041
	R^2	0.995

Table 5.6 (continued)

Isotherm	Parameters	Value
Toth	$q_m{}^o$, mg/g	49.93
	δ, 1/K	0.0031
	$K_T{}^o$	0.0171
	E, kJ/mol	8.09
	n_o	1.074
	Γ	0.0040
	R^2	0.993

Figures 5.13–5.15 clearly show that the Langmuir, Sips, and Toth models can represent the adsorption experimental data of ammonium onto modified bentonite fairly well. The comparisons on the applicability of these isotherms based on the physical meaning of the fitted parameters are given in Tables 5.5 and 5.6. The modification of bentonite using NaOH increased its adsorption capacity toward the ammonium ion as shown in Figure 5.13. This phenomenon was captured by $q_m{}^o$. The value of $q_m{}^o$ for modified bentonite is higher than the pristine bentonite.

The next parameter to be assessed is the expansion coefficient δ, which is essentially independent on the type of adsorbent. The values of δ obtained from the fitting of experimental data for both bentonite and its modified form are essentially the same as given in Tables 5.5 and 5.6. Since the modified bentonite has higher adsorption capacity than the pristine bentonite, ammonium ions are more strongly attached to the surface of the modified bentonite than the pristine one and the surface-modified bentonite is covered by a greater number of ammonium ions. The higher value of affinity coefficient of modified bentonite than pristine bentonite is an evidence to support the previous statement.

The modification of bentonite using NaOH and microwave irradiation increased the heterogeneity of the system as well, and this phenomenon is also captured by the parameter n_o in Sips and Toth models. The values of n_o for both equations obtained from the fitting of the ammonium adsorption experimental data onto modified bentonite are higher than pristine bentonite indicating that the system is more heterogeneous after the adsorption. This phenomenon is also supported by the value of fitted heat of adsorption parameter. Since the system is more heterogeneous, the interaction force between adsorbent and adsorbate becomes stronger and more heat is released during the adsorption process as indicated in Table 5.6. All these evidences conclude that Langmuir, Sips, and Toth models could represent the adsorption experimental data well for both adsorption systems.

5.6 Adsorption of total ammonia nitrogen from aquaculture, a case study

Adsorption of TAN from the real aquaculture system was also conducted using bentonite and NaOH-modified bentonite. All experiments were carried out in continuous and circulation mode. The fiberglass fish tank used in the study has a capacity of 750 L and was filled with 500 L of tap water. The fish tank contains 30 Japanese Koi fishes with an average length of 30 cm and an average weight of 1 kg. Resun submerged water pump with the capacity of 4,000 L/h was used to circulate water in the fish tank. The protein level in the feed was 30% (crude protein) and the feeding rate was 3 times daily, with a total food intake of 1% of the fish body weight daily. Prior to the adsorption experiment, the total ammonia concentration in the fish tank was measured every 60 min to obtain the profile of ammonia concentration. The measurement was begun after 60 min of fish feeding. The pH of the system was maintained around 6.5. The physical characteristics of the water used for the fish tank are given in Table 5.7.

Table 5.7: Characteristics of water for fish tank.

Parameter	Value
pH	7.04
Total alkalinity, mg/L as $CaCO_3$	58
Total hardness, mg/L as $CaCO_3$	79
Fe, mg/L	0.04
Mn, mg/L	0.01
F, mg/L	–
Cl, mg/L	84
SO_4, mg/L	78
NO_3, mg/L	–
Total ammonia, mg/L	–
PO_4, mg/L	–

The concentration value reported in Figure 5.16 is the total ammonia (both unionized ammonia and ammonium ion). It shows the ammonia concentration in the fish tank as a function of time before the adsorption experiment. The main source of ammonia in the fish tank is fish excretion. The rate of ammonia excretion directly depends on

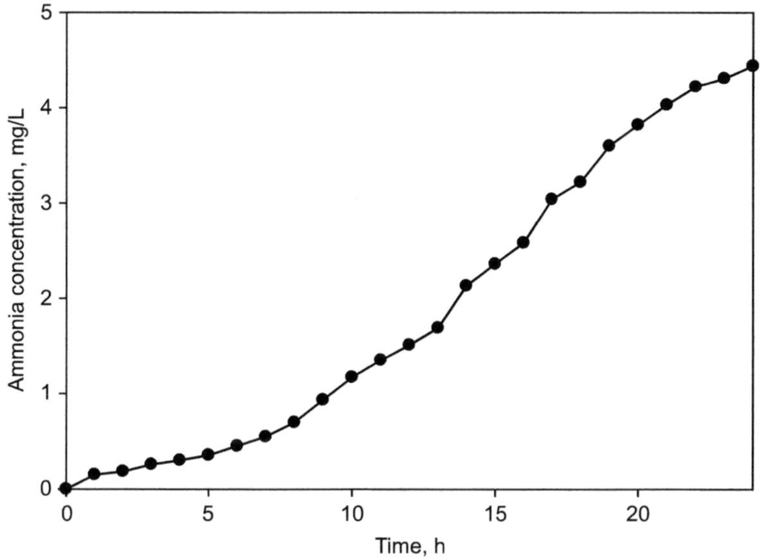

Figure 5.16: Ammonia concentration in fish tank as a function of time.

feeding rate and the protein level in the feed [79]. Without any treatment, the level of ammonia in the fish tank increases with time as shown in Figure 5.16. As dietary protein is consumed and broken down in the body, some nitrogen is used for growth, some used as energy, and the rest is excreted from the fish as ammonia [79]. Since the Koi fishes used in this study consumed a significant amount of protein (around 3.0 g crude protein daily), the excreted ammonia in water also increases significantly.

After 24 h, the concentration of total ammonia in the fish tank has reached 4.44 mg/L. At this level of concentration, most Koi fishes suffer from ammonia poisoning and appear sluggish, and often swim at the surface as if gasping for air. For the removal of ammonia, water in the fish tank was circulated through a fixed bed column containing 5 kg of adsorbent (bentonite or modified bentonite). During the removal of ammonia, the fish feeding was stopped in order to prevent further excretion of ammonia into the water. The behaviors of the Koi fish as well as the total ammonia concentration in the fish tank were carefully monitored in order to avoid further damage to the Koi fish. The concentration of total ammonia in the system was measured every 3 h, and the removal efficiency of each adsorbent (bentonite and modified bentonite) is given in Figure 5.17. This figure clearly shows that the modified bentonite has better removal efficiency than the pristine one. This result confirms that the bentonite and its modified form have the capability to remove ammonia from aquaculture system.

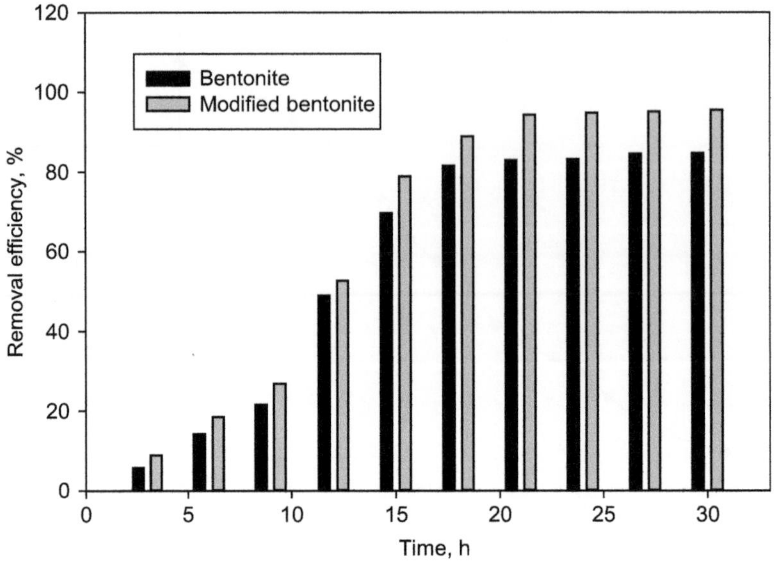

Figure 5.17: Removal of total ammonia from fish tank using bentonite and NaOH-modified bentonite.

5.7 Future direction of the application of bentonite for aquaculture industry

High adsorption capability of bentonite has found a wide range of industrial applications. Currently, the main applications of bentonite in industrial scale are as drilling fluids, binder, adsorbent, and so on. In the frying oil industry, bentonite is used to bleach oil [43]. As supplements for animal, bentonite acts as gut protectants which rapidly and preferentially bind aflatoxin from the digestive tract and thus reduce their adsorption into the animal [80].

In the aquaculture industry, bentonite (sodium bentonite) is usually used for the construction, repairing pond leaks, and excessive water seepage. Sodium-type bentonite has the ability to swell until 20 times of its original volume by absorbing water. This volume expansion allows bentonite to plugged pores in soils and prevents water seepage. Another application of bentonite in small-scale aquaculture industry is to prevent and control microalgae bloom [81, 82].

Compared to zeolites that have been widely used to remove and control ammonia from aquatic environment, bentonite has comparable adsorption capacity and even has the higher adsorption capability than several types of zeolites as listed in Table 5.1. The adsorption experimental data of total ammonia onto bentonite and its modified form reveal that this natural adsorbent has excellent adsorption capacity. The experiment in the real aquaculture system also indicates that this clay material possesses good removal efficiency toward TAN.

Several important characteristics of bentonite and its modified forms make this clay a valuable material as a large-scale adsorbent for aquaculture:
- High adsorption capacity
- Adsorption capacity can be easily improved through chemical/physical modification
- Available in large quantities
- Easy regeneration

There is no doubt that bentonite has high adsorption capacity toward ammonium ions as demonstrated in our experiments. However, since the adsorption capacity of bentonite is strongly influenced by its chemical composition, variation of adsorption ability especially for natural bentonite still cannot be avoided. Yet bentonites from several locations that have low natural binding capacity still widely exist.

Their adsorption capacity normally can be improved by pretreatment or modification using physical or chemical methods. Chemically, modification is usually performed by adding chemicals such as NaOH, organic surfactants, and acid. Physical modifications are usually performed by heat, microwave irradiation, boiling, and so on. To implement bentonite as adsorbents for aquaculture industry further testing in large scale as well as economic analysis should be conducted.

5.8 Conclusions

The use of clay materials for the removal of ammonia from aqueous solution has been widely explored, and most studies employed zeolites as the adsorbent for that purpose. Although bentonite has high adsorption capability in removing ammonia from aqueous solution, the use of this clay mineral to adsorb ammonia from water environment is still scarce. In this chapter, the potential application of bentonite as adsorbent for the removal of ammonia from aqueous solution as well as in the real aquaculture system is presented. The adsorption experimental data of total ammonia onto bentonite and its modified form reveal that this natural adsorbent has excellent adsorption capacity toward TAN. The Langmuir, Sips, and Toth equations can represent the adsorption experimental data of ammonia onto bentonite and its modified forms fairly well.

Acknowledgments: The authors acknowledge the financial support from Directorate General of Higher Education, Indonesia Ministry of Cultural and Education through Competency Research Grant with the contract number 003/SP2H/P/K7/KM/2015. The authors thank Dr. David Barkley from N. Simonson & Company (http://www.virtlab.com) for assistance in language editing.

References

[1] Gu, D., Zhu, X., Vongsay, T., Huang, M., Song, L., and He, Y., *Polish Journal of Environmental Studies*, 2013, 22, 13–49.

[2] Francis-Floyd, R., Watson, C., Petty, D., and Pounder, D.B., *Fisheries and Aquatic Sciences FA16*, University of Florida, 2012. http://edis.ifas.ufl.edu

[3] Ammonia in the aquatic environment, Canadian Environmental Protection Act, 1999.

[4] Huang, H., Xiao, D., Pang, R., Han, C., and Ding, L., *Chemical Engineering Journal*, 2014, 256, 431.

[5] Cooper, P., Day, M., and Thomas, V., *Journal of Institute Water Environment and Management*, 1994, 8, 84.

[6] Emaranon, E., Ulmanu, M., Fernandez, Y., Anger, I., and Castrillon, L., *Journal of Hazardous Material*, 2006, 137, 1402.

[7] Alshameri, A., Ibrahim, A., Assabri, A.M., Lei, X., Wang, H., and Yan, C., *Powder Technology*, 2014, 258, 20.

[8] Alshameri, A., Yasir-Al-Ani, C.Y., Dawood, A.S., Ibrahim, A., and Zhou, C., *Journal of The Taiwan Institute of Chemical Engineers*, 2013, 45, 554.

[9] Li, M., Zhu, X., Zhu, F., Ren, G., Gao, G., and Song, L., *Desalination*, 2011, 271, 295.

[10] Huang, H., Xiao, X., Yan, B., and Yang, L., *Journal of Hazardous Materials*, 2010, 175, 247.

[11] Zhang, L., and Jahng, D., *Journal of Hazardous Materials*, 2010, 182, 536.

[12] Huang, H., Chen, Y., Jiang,Y., and Ding, L., *Chemical Engineering Journal*, 2014, 254, 418.

[13] Tice, R.C., and Kim, Y., *Water Research*, 2014, 64, 61.

[14] Hutardo, C.F., and Cancino-Madariaga, B., *Aquaculture Engineering*, 2014, 58, 29.

[15] Kurniawan, A., Kosasih, A.N., Febrianto, J., Ju, Y.H., Sunarso, J., Indraswati, N., and Ismadji, S., *Chemical Engineering Journal*, 2011, 172, 158.

[16] Jha, V.K., and Hayashi, S., *Journal of Hazardous Materials*, 2009, 169, 29.

[17] Yusof, A.M., Keat, L.K., Ibrahim, Z., Majid, Z.A., and Nizam, N.A., *Journal of Hazardous Materials*, 2010, 174, 380.

[18] Siljeg, M., Foglar, L., and Kukucka, M., *Journal of Hazardous Materials*, 2010, 178, 572.

[19] Karadag, D., Koc, Y., Turan, M., and Armagan, B., *Journal of Hazardous Materials*, 2006, 136, 604.

[20] Arslan, A., and Veli, S., *Journal of The Taiwan Institute of Chemical Engineers*, 2012, 43, 393.

[21] Lei, L., Li, X., and Zhang, X., *Separation and Purification Technology*, 2008, 58, 359.

[22] Weatherly, L.R., and Miladinovic, N.D., *Water Research*, 2004, 38, 4305.

[23] Wang, Y., Kmiya, Y., and Okuhara, T., *Water Research*, 2007, 41, 269.

[24] Malekian, R., Abedi-Koupai, J., Eslamian, S.S., Mousavi, S.F., Abbaspour, K.C., and Afyuni, M., *Applied Clay Science*, 2011, 51, 323.

[25] Zhou, L., and Boyd, C.E., *Aquaculture*, 2014, 432, 252.

[26] Moussavi, G., Talebi, S., Farrokhi, M., and Sabouti, R.M., *Chemical Engineering Journal*, 2011, 171, 1159.

[27] Halim, A.A., Aziz, H.A., Johari, M.A.M., and Ariffin, K.S., *Desalination*, 2010, 262, 31.

[28] Zhang, M., Zhang, H., Xu, D., Han, L., Niu, D., Tian, B., Zhang, J., Zhang, L., and Wu, W., *Desalination*, 2011, 271, 111.

[29] Li, M., Zhu, X., Zhu, F., Ren, G., Cao, G., and Song, L., *Desalination*, 2011, 271, 295.

[30] Zhang, M., Zhang, H., Xu, D., Han, L., Niu, D., Zhang, L., Wu, W., and Tian, B., *Desalination*, 2011, 277, 46.

[31] Leyva-Ramos, R., Monsivais-Rocha, J.E., Aragon-Pina, A., Berber-Mendoza, M.S., Guerrero-Coronado, R.M., Alonso-Davila, P., and Mendoza-Baron, J., *Journal Environment Management*, 2010, 91, 2662.

[32] Saltali, K., Sari, A., and Aydin, M., *Journal of Hazardous Material*, 2007, 141, 258.
[33] Pabalan, R.T., and Bertetti, F.P. Cation-Exchange Properties of Natural Zeolites. *Reviews in Mineralogy and Geochemistry*, 2001, 45(1), 453–518.
[34] Febrianto, J., Kosasih, A.N., Sunarso, J., Ju, Y.H., Indraswati, N., and Ismadji, S., *Journal of Hazardous Materials*, 2009, 162, 616.
[35] Wang, Y.J., Cannon, F.S., Komarneni, S., and Voigt, R.C., *Environment Science and Technology*, 2005, 39, 7712.
[36] Maiti, M., and Bhowmick, A.K., *Composites Science Technology*, 2008, 1, 1.
[37] Koyuncu, H., *Applied Clay Science*, 2008, 38, 279.
[38] Khenifi, A., Bouberka, Z., Sekrane, F., Kameche, M., and Derriche, Z., *Adsorption*, 2007, 13, 149.
[39] Richards, S., and Bouazza, A., *Applied Clay Science*, 2007, 37, 133.
[40] Sanchez-Martin, M.J., Rodriguez-Cruz, M.S., Andrades, M.S., and Sanchez-Camazano, M., *Applied Clay Science*, 2006, 31, 216.
[41] Mostafa, B.A., Assaad, F.F., and Attia, M., *Journal of Applied Pollution Science*, 2007, 104, 1496.
[42] Catarino, S., Madeira, M., Monteiro, F., Rocha, F., Curvelo-Garcia, A.S., and de Sousa, R.B., *Journal of Agricultural and Food Chemistry*, 2008, 56, 158.
[43] Tanjaya, A., Indraswati, Sudono, N., and Ismadji, S., *Indonesian Journal of Chemical Engineering*, 2006, 5, 429.
[44] Salazar, F.N., Achaerandio, I., Labbe, M.A., Guell, C., and Lopez, F., *Journal of Agricultural Food Chemistry*, 2006, 54, 9955.
[45] Magana, S.M., Quintana, P., Aguilar, D.H., Toledo, J.A., Angeles-Chavez, C., Cortes, M.A., Leon, L., Freile-Pelegrin, Y., Lopez, T., and Sanchez, R.M.T., *Journal of Molecular Catalyst A – Chemistry*, 2008, 281, 192.
[46] Elmore, A.R., Andersen, F.A., *International Journal of Toxicology*, 2003, 22, 37.
[47] Aguzzi, C., Cerezo, P., Viseras, C., and Caramella, C., *Applied Clay Science*, 2007, 36, 22.
[48] Viseras, C., Aguzzi, C., Cerezo, P., and Lopez-Galindo, A., *Applied Clay Science*, 2007, 36, 37.
[49] Borutskii, P.N., Kozlova, E.G., Podkletnova, N.M., Gil'chenok, N.D., Sokolov, B.G., Zuev, V.A., and Shatovkin, A.A., *Petroleum Chemistry*, 2007, 47, 250.
[50] Soetaredjo, F.E., Ayucitra, A., Ismadji, S., and Maukar, A.L., *Applied Clay Science*, 2011, 53, 341.
[51] Juang, R.S., Lin, S.-H., and Tsao, K.-H., *Journal of Colloid Interface Science*, 2002, 254, 234.
[52] Koyuncu, H., *Applied Clay Science*, 2008, 38, 279.
[53] Anggraini, M., Kurniawan, A., Ong, L.K., Martin, M.A., Liu, J.C., Soetaredjo, F.E., Indraswati, N., and Ismadji, S., *RSC Advances*, 2014, 4, 16298.
[54] Ong, L.K., Soetaredjo, F.E., Kurniawan, A., Ayucitra, A., Liu, J.C., and Ismadji, S., *Chemical Engineering Journal*, 2014, 241, 9.
[55] Chandra, I.K., Ju, Y.H., Ayucitra, A., and Ismadji, S., *International Journal Environment Science Technology*, 2013, 10, 359.
[56] Kurniawan, A., Sutiono, H., Indraswati, N., and Ismadji, S., *Chemical Engineering Journal*, 2013, 189–190, 264.
[57] Maramis,Sandy, V., Kurniawan, A., Ayucitra, A., Sunarso, J., and Ismadji, S., *Frontier Chemical Science and Engineering*, 2012, 6, 58.
[58] Nathaniel, E., Kurniawan, A., Soetaredjo, F.E., and S. Ismadji, *Desalination and Water Treatment*, 2011, 36, 280.
[59] Rahardjo, A.K., Susanto, M.J.J., Kurniawan,A., Indraswati, N. and Ismadji, S., *Journal Hazardous Materials*, 2011, 190, 1001.

[60] Kurniawan, A. , Sutiono, H., Ju, Y.H., Soetaredjo, F.E., Ayucitra, A., Yudha, A., and Ismadji, S., *Microporous Mesoporous Materials*, 2011, 142, 184.

[61] Erdemoglu, M., Erdemoglu, S., Sayilkan, F., Akarsu, M., Sener, S., and Sayilkan, H., *Applied clay science*, 2004, 27, 41.

[62] Yesi, F.P. Sisnandy, Y.H. Ju, Soetaredjo, F.E., and Ismadji, S., *Adsorption Science and Technology*, 2010, 28, 847.

[63] Koswojo, R., Utomo, R.P., Ju, Y.H., Ayucitra, A., Soetaredjo, F.E., Sunarso, J., and Ismadji, S., *Applied Clay Science*, 2010, 48, 81.

[64] Putra, E.K., Pranowo, R., Sunarso, J., Indraswati, N., and Ismadji, S., *Water Research*, 2009, 43, 2419.

[65] Zamparas, M., Drosos, M., Georgiou,Y., Deligiannakis, Y., and Zacharias, I., *Chemical Engineering Journal*, 2013, 225, 43.

[66] Cai, F.F., Yang, Z.H., Huang, J., Zeng, G.M., Wang, L., and Yang, J., *Journal of Hazardous Materials*, 2014, 275, 63.

[67] Venglovsky, J., Pacajova, Z., Sasakova, N., Vucemilo, M., and Tofant, A., *Veterinary and Medical Czech*, 1999, 44, 339.

[68] Bedelean, H., Măicăneanu, A., Burcă, S., and Stanca, M., *Hellenic Journal of Geoscience*, 2010, 45, 23.

[69] Abdullah, Syafalni, R., Abustan, I., and Ibrahim, A.N.M., *International Journal of Environmental Science*, 2013, 4, 379.

[70] Buragohain, P., Sredeep, S., and Saiyouri, N., *International Journal of Chemical Environment and Biological Science*, 2013, 1, 157.

[71] Shao, H., Zhang, D., Zhang,Z., and Wang, D., *Advanced Journal of Food Science and Technology*, 2013, 5, 721.

[72] Órfão, J.J.M., Silva, A.I.M., Pereira, J.C.V., Barata, S.A., Fonseca, I.M., Faria, P.C.C., and Pereira, M.F.R., *Journal of Colloid and Interface Science*, 2006, 296, 480.

[73] Patel,H.A., Somani,R.S., Bajaj,H.C., and Jasra,R.V., *Current Science*, 2007, 92, 1004.

[74] Do, D.D. Adsorption analysis: equilibria and kinetics. Imperial College Press, UK, 1998.

[75] APHA, Standard methods for the examination of water and wastewater, 21st ed. APHA, AWWA, WPCF, Washington DC, USA, 2005.

[76] Thornton,A., Pearce,P., and Parsons,S.A., *Journal of Hazardous Materials*, 2007, 147, 883.

[77] Ismadji, S. and Bhatia, S.K., *Langmuir*, 2001, 17, 1488.

[78] Lide, D.R., and Frederiske, H. P. R., *CRCHandbookof ChemistryandPhysics*; CRC Press: New York, 1996.

[79] Hargreaves, J.A. and Tucker, C.S., *SRAC Publication* No. 4603 2004.

[80] Trckova, M., Matlova, L., Dvorska, L., and Pavlik, I., *Veterinary Medical Czech*, 2004, 49, 389.

[81] Padilla, L.V., McGlone, M.L.S.D., and Azanza, R.V., *Science Diliman*, 2006, 18, 35.

[82] Sengco, M. and Anderson, D., *Journal of Eukaryotic Microbiology*, 2004, 51, 169.

Ayşenur Sağlam, Sema Bektaş, and Adil Denizli

6 Ion-imprinted thermosensitive macroporous cryogels for cadmium removal

6.1 Introduction

Environmental stimuli-responsive polymers are getting more and more attention from both technological and scientific aspects, because of their promising application potentials in the fields of controlled drug delivery [1], enzyme immobilization [2], chemical separation [3], catalysis [4, 5], sensor [6], and so on. Environmental stimuli-responsive polymers are also called smart polymers or environmentally sensitive polymers. Smart polymers are soluble, surface-coated, or cross-linked polymers that exhibit relatively large and sharp physical or chemical changes in response to small physical or chemical external stimuli such as temperature or pH [7, 8]. These polymers can respond the specific environmental stimuli by changing their size.

In literature to date, limited number of study reported novel adsorbents using thermosensitive hydrogels for trapping heavy metal ions [9–12], where pNIPA was used as a thermosensitive polymeric backbone. A chelating group, which interacts with heavy metal ions, was introduced into pNIPA meanwhile applying molecular imprinting technique [12] using a specific metal as the template. The molecularly imprinted adsorbents reconstruct multipoint adsorption sites at a specific temperature and disrupt them through swelling/deswelling deformation at another specific temperature. Such adsorbents, therefore, are suitable for the control of adsorption and desorption of a specific heavy metal ion with respect to temperature variation. Using the adsorbents described here provides an energy-saving and environmentally friendly process for the separation of both undesirable and valuable metals in aquatic environment, industrial effluents, and so on.

In this study, ion-imprinted p(NIPA-MAC), poly(*N*-isopropylacrylamide-*N*-methacryloyl-L-cysteine), and (MIP–Cd(II)) thermosensitive hydrogels were prepared for removing Cd(II) ions from aqueous media selectively. MIP–Cd(II) hydrogel was synthesized by free radical polymerization technique. The effects of cross-linker and MAC monomer amount used in the synthesis of the hydrogel were investigated in order to improve the swelling/deswelling behavior of the hydrogel. For comparison, pNIPA and nonimprinted p(NIPA-MAC) (NIP) hydrogels were also prepared. The characterization of the thermosensitive hydrogels was carried out by using swelling test, FTIR (Fourier transform infrared), elemental analysis, scanning electron microscopy (SEM), and energy-dispersive x-ray spectrometry (EDX) techniques. In the second part

Ayşenur Sağlam, Sema Bektaş, Adil Denizli, Hacettepe University, Department of Chemistry, Beytepe, Ankara, Turkey

https://doi.org/10.1515/9783110650600-006

of the study, adsorption and desorption of Cd(II) ions from aqueous solutions onto the MIP–Cd(II), NIP, and pNIPA hydrogels was investigated in batch processes. The effects of the initial Cd(II) concentration and pH on the adsorption rate and adsorption capacity were studied for both NIP and MIP–Cd(II) hydrogels. Selectivity of the MIP–Cd(II) hydrogels for Cd(II) ions was investigated by using interfering metal ion mixtures of Pb(II), Cu(II), Cr(III), and Fe(III). Repeated use of the MIP–Cd(II) hydrogels in aqueous solutions and the recovery of Cd(II) ions from a certified water sample were also studied.

6.2 Materials and methods

6.2.1 Materials

N-Isopropylacrylamide (NIPA), *N,N*-methylenebis(acrylamide) (MBAA), ammonium peroxodisulfate (APS), *N,N,N',N'*-tetramethylenediamine (TEMED) were obtained from Aldrich Chem. Co. (USA) and used as supplied without further purification. $Cd(NO_3)_2.4(H_2O)$ was of reagent grade and purchased from Sigma (St Louis, USA), respectively. The functional monomer, *N*-methacryloyl-L-cysteine (MAC) was supplied from Nanoreg Ltd. Şti. (Ankara, Turkey) and used as received. Deionized water with 18.2 µS of specific conductivity was obtained from a Milli-Q water purification system (Millipore) used in all experiments.

6.2.2 Preparation of polymeric hydrogels

6.2.2.1 Prepolymer complex of MAC with Cd(II) ions

The chelating functional monomer, MAC, was supplied from Nanoreg Ltd Şti, which was prepared according to the method developed by Denizli et al. [13]. In order to prepare MAC–Cd(II) complex, solid MAC (0.120 g) was dissolved in 10 mL of deionized water and cadmium nitrate (0.090 g) was added to this solution at room temperature while stirring with a magnetic stirrer for 3 h. Then the aqueous metal–monomer prepolymer complex was used in the hydrogel synthesis without any further treatment.

6.2.2.2 Preparation of MIP–Cd(II) hydrogels

MIP–Cd(II) hydrogels were synthesized by free radical polymerization of NIPA and MAC–Cd(II) complex in aqueous solutions. MBAA and APS–TEMED were used as a cross-linker agent and the redox initiator couple, respectively. First, NIPA (0.85 g) was added slowly into MAC–Cd(II) complex while stirring the solution magnetically at 300 rpm. After that, APS (10 mg) was added and stirred for 1 h. Then, the MBAA (20 mg) was added to the solution. After that, the solution was purged with nitrogen gas for 10 min. Finally, TEMED (10 μL) was added and the polymer solution was placed in glass tubes of 6 mm inner diameter and 4 cm long. The glass tubes were sealed and the polymerization was carried out at room temperature for 24 h. After gelation was completed, all the gels were taken out of the glass tubes and washed consecutively with deionized water by incubating the gels in cold water (swollen form) and hot water (shrunken form) alternatively to remove the residual monomers and the initiator. The hydrogels were dried in a lyophilizator (Christ, Alpha LD 1–2 Plus, Germany).

Similarly, NIP and pNIPA hydrogels were also prepared. In the preparation of NIP gel, the chelating monomer (0.120 g of MAC) without complexing with Cd(II) ions was included into the polymerization recipe. pNIPA gel was prepared without the chelating monomer.

In order to remove Cd(II) ions, freeze–dried MIP–Cd(II) hydrogels were immersed into 50 mL of HNO_3 solution (0.1 M) at room temperature for 6 h. This procedure was continued until all Cd(II) ions were removed from the hydrogels, which checked by measuring Cd(II) amount in washing solution. After that the template-free polymers were dried in a lyophilizator.

6.2.3 Characterization of the hydrogels

6.2.3.1 Temperature dependence of swelling ratios

Hydrogel samples were equilibrated in deionized water in the temperature range of 5.0–60.0 °C. This range was selected for the evaluation of the effect of the temperature on swelling behavior below and above LCST of the NIPA. The swelling experiments were performed in deionized water for 24 h while adjusting temperature (±0.1 °C) in a water bath with a cooling unit (Julabo, F34, Germany). After 24 h interaction with deionized water at a predetermined temperature, the hydrogels were taken out from water, wiped out with a filter paper, and weighed until constant weight. After whole swelling experiments, the hydrogels were dried in a vacuum oven at 40 °C for 3 days to determine dry weight of each sample. Each experiment was repeated three times, and the average values were used for determining the swelling degree (SD) by applying the following equation:

$$SD = (W_{swollen} - W_{dried}) / W_{dried} \tag{6.1}$$

where $W_{swollen}$ and W_{dried} are the weights of the swollen and dried samples, respectively. The LCST of the hydrogel samples was determined as the abscissa of the inflection point of the swelling ratio versus temperature curves.

6.2.3.2 Swelling rate

The swelling rates of the hydrogels were evaluated at 22 °C with gravimetrical measurements. The hydrogel samples dried under vacuum at 40 °C were interacted with deionized water at constant temperature while weighted at regular time intervals. Water uptake was calculated using the following equation:

$$\text{Water uptake} = \left[\left(W_{\text{time}} - W_{\text{dried}}\right) / \left(W_{\text{eq}} - W_{\text{dried}}\right)\right] \times 100 \tag{6.2}$$

where W_{time}, W_{eq}, and W_{dried} are the weights of hydrogel samples for swollen at time t, equilibrium, and dried state. It should be noted that the water attached on the hydrogel surface was wiped out with a filter paper.

6.2.3.3 FTIR characterization

The chemical structures of pNIPA, NIP, and MIP–Cd(II) hydrogels were characterized by using a Fourier transform infrared (FTIR) spectrometer (FTIR 8,000 Series, Shimadzu, Japan) in the wavenumber range of 4,000–400 cm^{-1}. The KBr disk technique was performed by mixing 2 mg of hydrogel sample with 98 mg of KBr (IR-Grade). Hydrogel specimens were dried with a freeze-drying method.

6.2.3.4 Morphologies of hydrogels

The surface morphology of pNIPA, NIP, and MIP–Cd(II) hydrogel samples were evaluated by SEM (Quanta 400-ESEM, FEI). First, these samples were separately swollen in 50 mL of deionized water at two different temperatures (20 and 50 °C) until reaching to equilibrium state. Then, liquid nitrogen was used for quick freezing of these samples and a freeze-dryer (Christ, Alpha LD 1–2 Plus, Germany) was used for lyophilizing them at −45 °C and 20 mbar for 2 days. The lyophilized samples were then cut off carefully and SEM images were determined by mounting them to the SEM holder and sputtering with thin gold film for 2 min. Finally, SEM images were taken at desired magnification.

6.2.3.5 Energy-dispersive x-ray analysis

The presence of cadmium ions in MIP–Cd(II) hydrogels was investigated by using an EDX in conjunction with SEM. Thus, the same hydrogel sample used for the SEM observation was analyzed. NIP sample, which served as a reference, was also investigated for its cadmium-free content.

6.2.4 Adsorption and desorption studies

6.2.4.1 Temperature-dependent adsorption studies

The Cd(II) adsorption properties of the MIP–Cd(II) hydrogel were determined batchwise. Cylindrical gel pieces (dry weight: 0.200 g) were added into a beaker containing 50 mL of aqueous solution of $Cd(NO_3)_2.4H_2O$ with various concentrations ranging from 0.100 to 100 mg/L, and at different pH values (in the range of 2.5–7.5) and then the beaker was sealed up and placed on a magnetic stirrer at a speed of 300 rpm at room temperature for 24 h. For the adsorption studies, the sealed beaker was placed in a thermostatic water bath shaker and operated under 300 rpm at designed temperature for 24 h. pH was adjusted by the addition of HNO_3 and NaOH. After the desired treatment periods, AAS (atomic absorption spectroscopy) was used for determining the concentration of the Cd(II) ions in the aqueous phase. The amount of Cd(II) adsorption on the MIP–Cd(II) hydrogels was evaluated by using appropriate mass balance. For comparison, similar adsorption experiments were also carried out with pNIPA and NIP hydrogels.

6.2.4.2 Desorption and reusability studies

After the adsorption was completed Cd(II)-adsorbed MIP–Cd(II) gels were first squeezed and the adsorption medium was placed in a thermostatic water bath at 50 °C for 24 h. At the end of this period, aqueous phase was removed from the hydrogels and its Cd(II) ions content was measured by using AAS.

The Cd(II) ions adsorbed on the hydrogels were almost completely desorbed from the hydrogels by means of acid treatment. The metal ion adsorbed hydrogel samples were immersed in 50 mL of HNO_3 solution (0.1 M) and stirred continuously at 300 rpm at room temperature for 3 h. The concentration of the Cd(II) ions in desorption medium was measured by AAS as mentioned earlier. The amount of Cd(II) ions adsorbed on hydrogels and desorbed into media were used to calculate desorption ratio, and reusability was evaluated by repeated adsorption–desorption experiments by using the same hydrogels in each cycle. About 50 mL of NaOH solution (50 mM) was used for regeneration of hydrogels after each desorption step.

6.2.5 Selectivity experiments

The competitive adsorptions of some metal ions were also studied to evaluate the ion-recognition behavior of the MIP–Cd(II) hydrogels. Pb(II), Cu(II), Cr(III), and Fe (III) ions were selected for selectivity evaluation. The adsorption experiments were carried out with 10.0 mg/L aqueous solutions of each metal ion and the mixture of these ions in the similar manner as described previously. Experiments were performed at a constant pH of 6.0 at 20 °C. The measurements of the concentrations of these metal ions were also performed with AAS.

6.3 Results and discussion

6.3.1 Preparation of hydrogels

6.3.1.1 Temperature dependence of swelling ratios of the hydrogels

The effect of the composition of hydrogels on their swelling abilities was evaluated in the temperature range of 5–60 °C (Figure 6.1). The swelling abilities of hydrogels were directly related to their MAC contents. The increase in MAC content from 0 to 120 mg caused the increase in swelling ratio from 32 to 34 °C. This increase that depends on the amino acid-based monomer MAC has a hydrophilic side chain, which improved the hydrophilicity of polymeric network. It is well known about structural property of NIPA-based polymers that they have a balance between their hydrophilic and hydrophobic segments by being controlled with many intermolecular and intramolecular interactions, such as hydrogen bonds and polymer–polymer interactions [14]. MAC segments enhanced the hydrophilic character of the polymeric network, which caused an increase in water incorporation ratio into polymeric network at temperature below LCST while shifting the balance toward hydrophilic

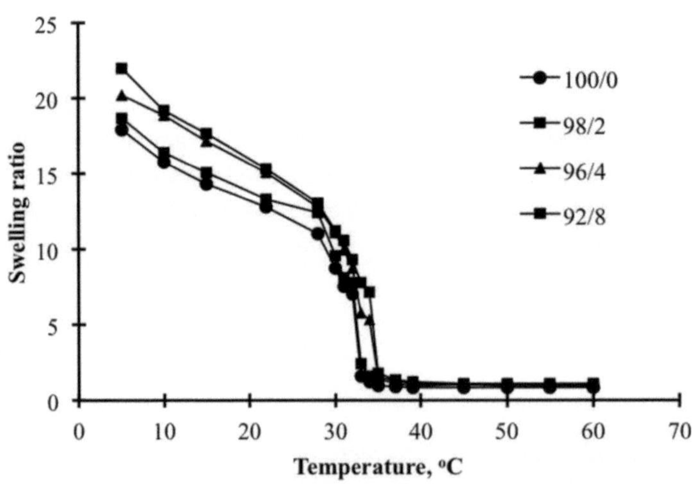

Figure 6.1: Influence of hydrogel composition on equilibrium swelling ratios (temperature base) of NIP hydrogels. The ratios are calculated on the molar base of monomers, NIPA and MAC.

polymer. In addition, MAC incorporation may also cause a decrease in molecular weight of polymeric chain. In the literature, the volume phase transition temperature or LCST of these types of hydrogels is defined as the temperature at which the swelling ratio has decreased to a half of its value at the initial temperature or room temperature [15]. The LCST value of hydrogels referred to the temperature at which the greatest phase separation was occurred; in other words, the highest change in temperature is based on the SD around the transition temperature ($\Delta SD/\Delta T$). It could be easily determined by the temperature at which most drastically decrease in swelling ratio from SD versus temperature curves [16, 17]. For further studies, NIP hydrogels with NIPA/MAC molar ratio of 92/8 was used (because of its higher metal ion adsorption capacity compared to other hydrogels with lower MAC content).

The effect of cross-linking density on the swelling ratio for hydrogels with the same NIPA/MAC monomer molar ratio of 92/8 and different amount of the MBAA cross-linker was also investigated. As shown in Figure 6.2, the hydrogels had the same LCST, 34 °C, with different amounts of the MBAA (1.0–4.0 wt%, based on total monomer). So the amount of the cross-linker did not influence the LCST and phase separation behavior of hydrogels evidently in the range of 1.0–4.0 wt%. As the cross-linking density within the hydrogel increased, swelling ratio difference did not change significantly at temperatures above LCST. At 10 or 20 °C (below LCST), however, for the hydrogels with higher cross-linking density, lower swelling ratio was observed, compared to hydrogels with lower cross-linking density. This result indicated that high cross-linking density made the structure collapse more tightly than in the other samples. It decreases the diffusion of chains and thus reduces the

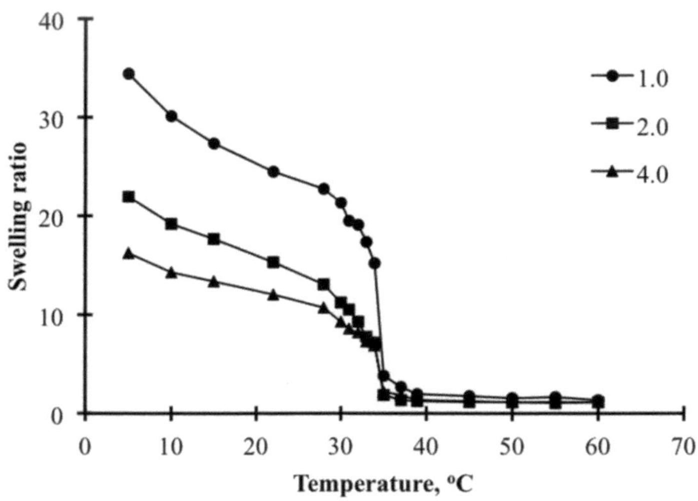

Figure 6.2: Influence of cross-linker content (wt%) on equilibrium swelling ratios (temperature base) of NIP hydrogels.

dilatation of three-dimensional structure of hydrogels. Hydrogel with cross-linking density of 2.0 wt% was chosen for further studies, because of its easy handling in the swollen form, compared to the one with 4.0 wt%.

Figure 6.3 shows the temperature-based swelling ratio and LCST of the pNIPA, NIP, and MIP–Cd(II) hydrogels in the temperature range of 5–60 °C. As shown in Figure 6.3, the swelling abilities of pNIPA, NIP, and MIP–Cd(II) hydrogels were contrary to temperature as expected. They swelled at lower temperature whereas shrinked at higher temperature because of the NIPA content. Under equilibrium swelling conditions, they showed an increase in swelling ability at lower temperatures, whereas they deswelled at high temperatures because of the aggregation of the network chains. The highest LCST value was observed at 34 °C for NIP gels, followed by MIP–Cd(II) and pNIPA hydrogels with the LCST of 33 and 32 °C, respectively.

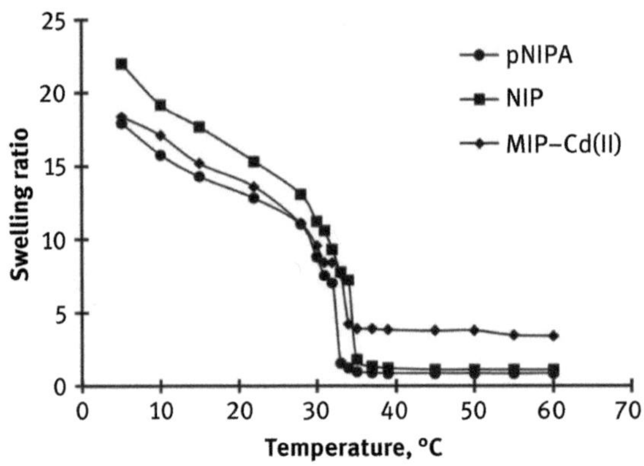

Figure 6.3: Equilibrium swelling ratios (temperature base) of p(NIPA), NIP, and MIP–Cd(II) hydrogels.

Although the swelling ratio of MIP–Cd(II) is somewhat lower than that of NIP, it is still higher than that of pNIPA at temperatures below LCST, as shown in Figure 6.4. As described earlier, incorporation of more hydrophilic monomer (MAC or MAC–Cd complex) to pNIPA hydrogels increases the LCST value because the ionized $-COO^-$ groups are sufficiently soluble to counteract the aggregation of the hydrophobic temperature-sensitive units. Also, the repulsion of the $-COO^-$ groups or the formation of hydrogen bonds between the amide groups in NIPA and the $-COO^-$ groups in MAC may impede the collapse induced by the NIPA components, increasing the LCST [18]. At temperatures above LCST, a notable result for the MIP–Cd(II) gel was observed. The data in Figure 6.3 were correlated to classical behavior of pNIPA-

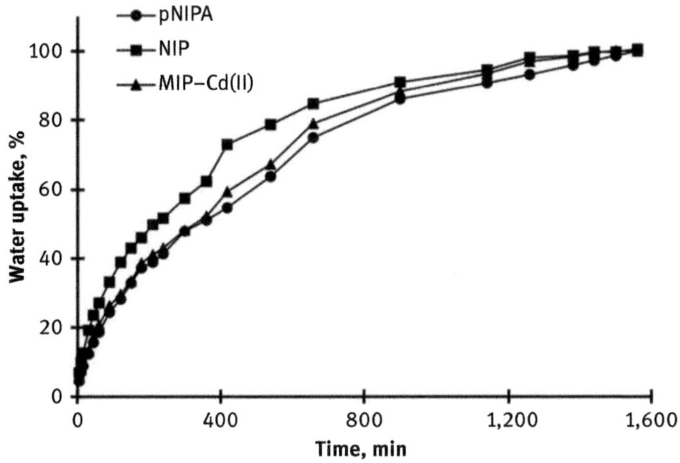

Figure 6.4: Swelling kinetics of the pNIPA, NIP, MIP–Cd(II) hydrogels at 22 °C.

based materials having a decrease in swelling ratio because of an increase in temperature. As mentioned earlier, MAC monomer (and/or MAC-Cd(II) complex) incorporation into hydrogel backbone unbalanced the equilibrium between hydrophilic and hydrophobic segments and shortened the polymeric chain, which decreased the magnitude of thermoinduced shrinkability of the hydrogels. The swelling ratio of MIP–Cd(II) hydrogel reduced from 17.1 to 3.4, as temperature varied from 10 to 60 °C, with a Δ swelling ratio of 13.7 ($SR_{10\,°C} - SR_{60\,°C}$). Over the same temperature range, Δ swelling ratios of pNIPA and NIP were 14.9 and 18.1, respectively. The decrease in the magnitude of Δ swelling ratio observed for MIP–Cd(II) hydrogel can be attributed to the structure of the MAC–Cd(II) complex influenced balance between the hydrophilic and hydrophobic units and morphology of the polymeric chain that prevented collapsing of the hydrogel above the LCST.

6.3.1.2 Swelling rates

Figure 6.4 shows the swelling kinetics of pNIPA, NIP, and MIP–Cd(II) hydrogels at room temperature (22 °C). During the swelling, in other words hydration process, the rate-limiting step is the penetration of water molecules into the pores of hydrogels. The data in Figure 6.4 show that the slope of the curve belonging to pNIPA is smaller than the others at the beginning of swelling process, which indicates that the hydration of pNIPA hydrogel was slower at the beginning of swelling process. In the case of pNIPA, about 8% of water was adsorbed within 15 min, about 18% within 60 min, and 40% at the end of 240 min. NIP hydrogel adsorbed 13% within first 15 min and 52% within 240 min, whereas MIP–Cd(II) hydrogel adsorbed 10%

and 44% within 15 and 240 min, respectively. This relationship suggests that faster diffusion of water molecules into the hydrogel network could be achieved if it had more hydrophilic (MAC monomer) contents.

Although MIP–Cd(II) hydrogel exhibits rather faster swelling rate with respect to pNIPA in Figure 6.4, it shows a slower swelling rate in comparison with NIP. This can indicate that incorporation of the MAC–Cd(II) complex into the hydrogel changed the systematic distribution of the hydrophilic monomer during the polymerization of the hydrogel when compared in the case of plain MAC used. So, the decrease in intramolecular interactions (hydrogen bonds) between MAC segments and NIPA groups would lead to a decrease in water absorption rate. In order to describe swelling process of hydrogels, three main controlling steps were suggested: diffusion, relaxation, and expansion. First, water molecules started to diffuse into hydrogel network. Then, polymer chain absorbed and relaxed subsequently. Finally, polymer network expanded through the surrounding media [19].

The structural stability during the swelling depends on porosity and wall thickness of hydrogels. Porosity enhanced the solvent diffusion into polymeric network, whereas thickness and rigidness of the polymeric walls diminished the relaxation of hydrated polymeric network. To evaluate the morphological changes during swelling we performed macroscopic observation by using SEM. As shown in Figure 6.5, the swelling rate was the macroscopic observation resulting from the combination of these three steps and the NIP hydrogel has a faster swelling rate than the pNIPA or the imprinted one.

6.3.1.3 FTIR characterization

The FTIR measurements of the dried hydrogels were also performed to characterize the chemical structure of those hydrogels (Supplementary File, Figure SF-1). From the given spectra we can find the amide I band (~1653 cm^{-1}) ascribed to the C = O stretch of PNIPA and the amide II band (~1546 cm^{-1}) due to the N–H bending vibration in every spectrum. The broad peak at the range from 3,200 to 3,600 cm^{-1} belongs to the N–H or O–H stretching vibration. Furthermore, the characteristic double peaks at 1,388 and 1,366 cm^{-1} for isopropyl group of NIPA appeared in all FTIR spectra of pNIPA, NIP, and MIP–Cd(II) hydrogels. Specifically, the appearance of the C–S stretching vibrations in the region of 700–600 cm^{-1} in both FTIR spectra of NIP and MIP–Cd(II) hydrogels denoted a successful incorporation of MAC groups into copolymeric hydrogels.

6.3.1.4 SEM observation of hydrogels

Morphology is a critical factor that determines response behaviors of pNIPA-based hydrogels [20]. The interior morphology of swollen (at 20 °C, below LCST) and

(a)
(b)
(c)
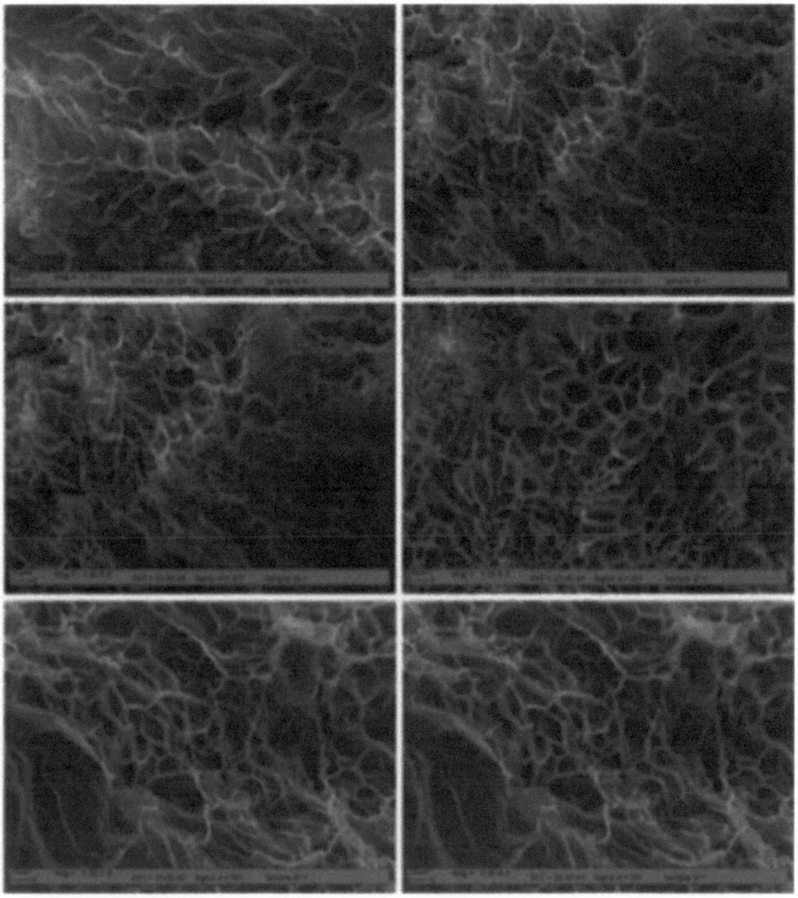

Figure 6.5: SEM images of freeze–dried (a) pNIPA, (b) NIP, and (c) NMIP–Cd(II) hydrogels swollen at 20 °C (at the left side) and shrunken at 50 °C (at right side). The size of the bar is 20 μm and the samples are viewed at magnification of 1.00 KX.

shrunken (at 50 °C, above LCST), and freeze–dried hydrogels is shown in Figure 6.5. As illustrated in Figure 6.5, among all the hydrogels in swollen form at 20 °C, the NIP hydrogel had the largest pore size with a >20 μm in diameter, and also exhibited more homogenous pore distribution like honeycomb. The enlarged porous network of the NIP hydrogel depends on the incorporation of MAC monomer having more polar hydrophilic character.

Amide groups of plain pNIPA chains make hydrogen bonds with water molecules while swelling in water at temperature below LCST [21]. The orientation of water molecules around hydrophobic groups forms a cage-like structure, which results in swelling of pNIPA hydrogels in water [22, 23]. With the introduction of MAC comonomer into NIPA backbone, both hydrogels [NIP and MIP–Cd(II)] have more complicated balance between hydrophilic and hydrophobic segments. Because of incorporation of sulfhydryl and other carboxylic acid group from MAC monomer, more water molecules participated in hydrogen bonds between polymeric chains and surrounding water molecules, which caused more ordered swelling kinetic and better organized structure around hydrogels. The results given in Figure 6.4 confirmed the higher swelling ratio of the NIP hydrogel.

According to the SEM images below LCST in Figure 6.5 (left side), moderately disorganized morphology and rather small pores observed for NIP hydrogel compared with the nonimprinted one may result from the molecular structure difference between the MAC and the MAC–Cd(II) complex. It can be suggested that with the introduction of MAC–Cd(II) complex during the polymerization, the propagation of pNIPA chains was assembled and/or cross-linked resulting in different orientations of the hydrophilic/hydrophobic groups (–COO^-, –NH, and isopropyl groups) between the complex monomer and the NIPA monomer due to the two pieces of MAC segments in each unit of the complex when compared with MAC. Finally, the resulting MIP–Cd(II) hydrogel exhibited a heterogeneously distributed matrix with a rather small pore from its NIP one. This observed morphological difference between the MIP–Cd(II) and NIP hydrogels has disclosed the lower swelling ratio of the MIP–Cd(II) compared to NIP hydrogel in Figure 6.5. As expected, the decrease in pore size causes a decrease in empty volume to absorb solvent molecules showing itself as lowering the swelling ratio [24–28].

As shown in Figure 6.5, the SEM images above LCST show that the pNIPA and NIP hydrogels exhibited rather compact network structure with small pores when compared with their corresponding SEM images below LCST. However, the MIP–Cd(II) hydrogel has dramatically changed porous matrix from open hollow (channel like) pores to round and shallow pores when the temperature changed from 20 to 50 °C. These results are consistent with the deswelling degree of the hydrogels. Zhang et al. reported that the hydrogel has appropriate releasing channels throughout the network, so they can shrink fast at temperature above LCST [24], besides, these water release channels should be kept open for the freed water to transfer out quickly and mostly (completely) as reported in literature [15]. Thus, the most possible reason that causing the MIP–Cd(II) hydrogel could exhibit lower deswelling at temperatures above LCST was incorporation of chelating functional monomer into hydrogel network which altering hydrogel composition and entrapping more water molecules in the interior pores; therefore, hydrogel resisted to deswell during the shrinking process.

6.3.1.5 EDX analysis

EDX, when combined with SEM, provides elemental analysis on areas as small as nanometers in diameter. The impact of the electron beam on the sample produces x-rays that are characteristic of the elements found in the sample. EDX spectra for NIP and MIP–Cd(II) hydrogels are given in Figure 6.6. From comparative analysis of the EDX spectra in Figure 6.6, the copolymerization of NIPA and MAC–Cd(II) complex was confirmed due to the appearance of the Cd peaks in the 3–4 keV range in the EDX spectrum of MIP–Cd(II) hydrogel but not in the spectrum of the NIP hydrogel. The feature peaks suggested a successful production of MIP–Cd(II) hydrogel.

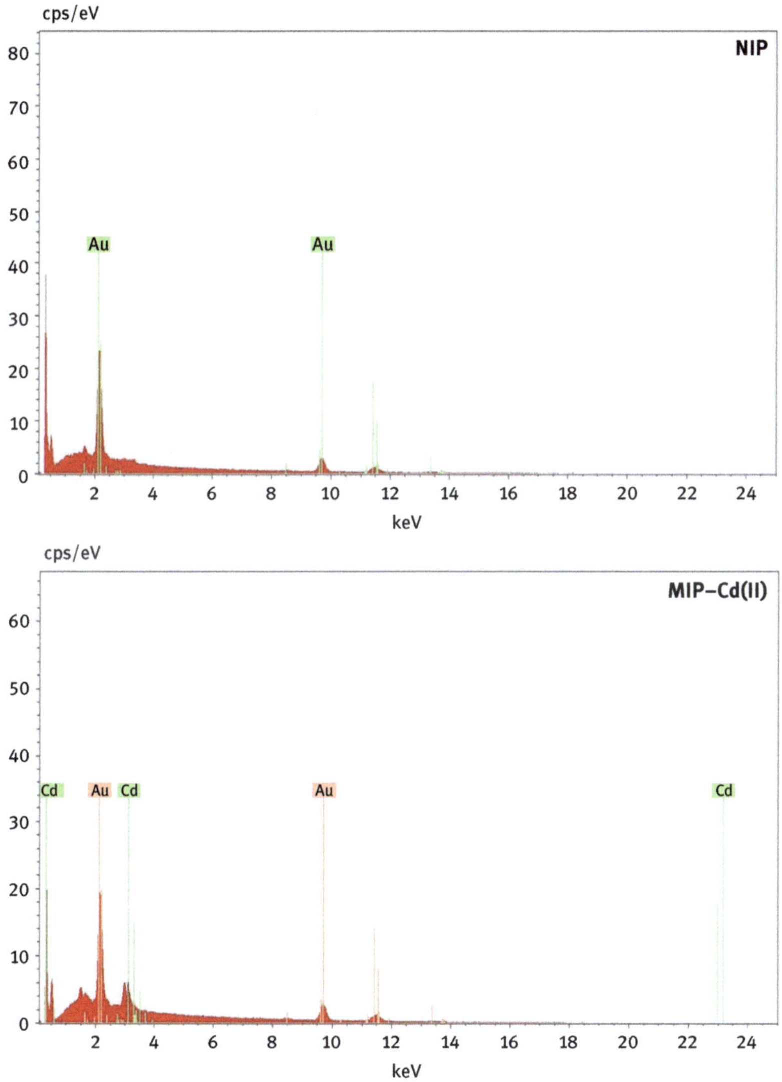

Figure 6.6: EDX spectra of NIP and MIP–Cd(II) hydrogels.

6.3.2 Adsorption studies

6.3.2.1 Adsorption rate

The variation of the Cd(II) ion adsorption amounts of the pNIPA, NIP, and MIP–Cd(II) hydrogels as a function of time is presented in Figure 6.7, and the adsorption conditions are given in the figure legend. All the hydrogels were previously shrunken at 50 °C, and then the adsorption was carried out at 20 °C. The initial slopes of these curves reflect the fast adsorption rates. High adsorption rates are observed at the beginning, and then plateau values (i.e., adsorption equilibrium) are gradually reached within 240 min for both NIP and MIP–Cd(II) hydrogels.

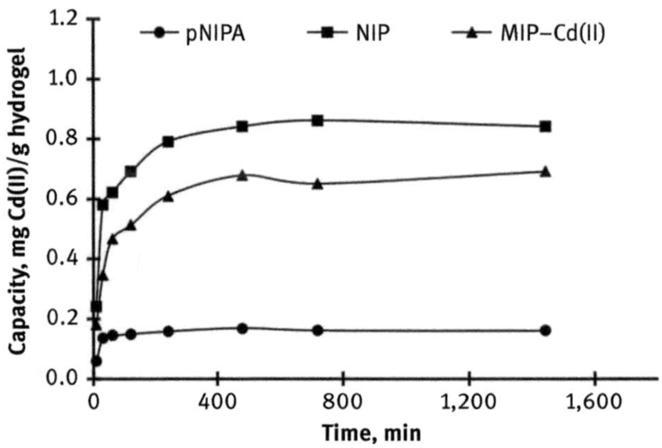

Figure 6.7: Time-dependent adsorption of Cd(II) ions on the pNIPA, NIP, and MIP–Cd(II) hydrogels. Adsorption conditions: 50 mL, 10 ppm Cd(II) solution; pH: 5.5; temperature 20 °C.

The slow adsorption is considered due to the fact that the swelling of the gel network requires some time course for the formation of the expanded network structure. In the case of pNIPA, equilibrium adsorption rate is observed within 60 min, which indicates that the Cd(II) ions adsorption is not affected with the expanded network structure of the pNIPA hydrogel compared to MAC incorporated hydrogels, due to poorer binding ability for Cd(II) ions of amide group in the NIPA.

In literature to date, several experimental data on the adsorption of various ions by thermosensitive polymers have shown a wide range of adsorption rates. Kanazawa et al. have studied adsorption/desorption properties of heavy metal ions by using poly(*N*-isopropyl acrylamide-co-*N*-(4-vinyl)benzyl ethylene diamine) [p (NIPA-Vb-EDA)] thermosensitive gels and they have found that adsorption equilibrium is reached in 100 h [12]. Tokuyama et al., using the same polymer but having different amount of cross-linker, have found the adsorption time as 1,100 min. Ju et al. studied the removal of Pb(II) ions from aqueous solutions by using p(NIPAM-co-benzo-18-crown-6-acrylamide) [p(NIPAM-co-BCAm)] hydrogels. They reported that the equilibrium adsorption time was 3.5 h [11]. According to these results both the NIP and the MIP–Cd(II) hydrogels have shown fast adsorption rates and are most probably due to high complexation of MAC monomer with Cd(II) ions and low diffusion barrier as a result of higher porous polymer network.

6.3.2.2 Effect of pH

The effect of pH on the Cd(II) adsorption using the NIP and the MIP–Cd(II) hydrogels is given in Figure 6.8. In all of the cases, the adsorption amount increased with increasing pH, reaching a maximum value at around pH 5.5. However, at low pH values, that is, below pH: 4.0, the adsorption amount is lower, which can be considered due to the fact that in such a low pH range, nitrogen, sulfur, and carboxyl ligands in MAC are strongly protonated.

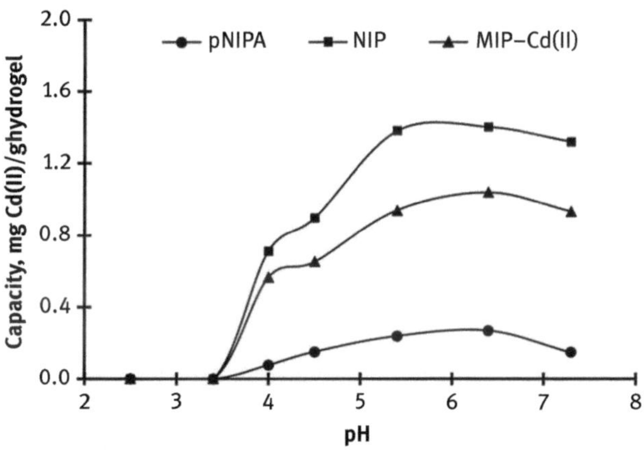

Figure 6.8: Effect of pH on the adsorption amount of pNIPA, NIP, and MIP–Cd(II) hydrogels for Cd(II) ions.

6.3.2.3 Adsorption capacity

In order to investigate the adsorption capacity of the MIP–Cd(II) hydrogels, 50 mL of Cd(II) solution at different initial concentrations ranging from 0.1 to 100 ppm (pH ~ 5.5) were interacted with 0.200 g hydrogel for 24 h, at 20 °C temperature. For comparison, adsorption capacities of the pNIPA and NIP were also studied by applying the same adsorption conditions. The Cd(II) ion adsorption amounts of all the hydrogels are given as a function of the initial concentration of Cd(II) ions within the aqueous adsorption medium in Figure 6.9.

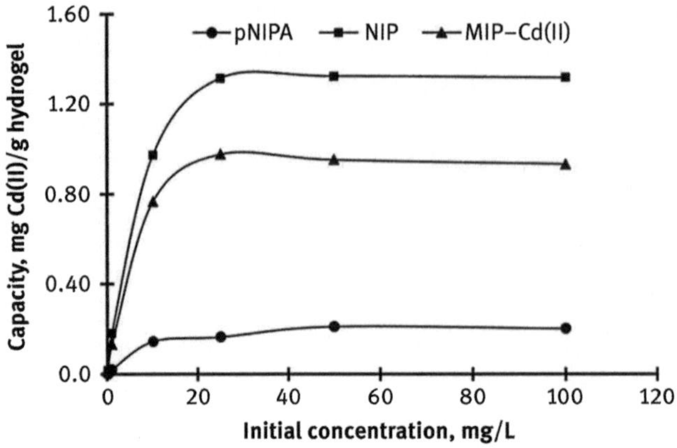

Figure 6.9: Effect of initial Cd(II) ion concentration on the adsorption amount of pNIPA, NIP, and MIP–Cd(II) hydrogels.

As shown in this figure, the amount of Cd(II) ions adsorbed per unit mass of the hydrogels increased with the initial concentration of Cd(II) ions, as expected. The adsorption capacities, in other terms saturation of the active sites (which are available for specific interaction with metal ions) on the hydrogels, of the MIP–Cd(II) and NIP hydrogels are 0.975 mg Cd(II)/g and 1.324 mg Cd(II)/g, respectively. The higher adsorption capacity of the NIP hydrogels than the imprinted one can be attributed to the higher MAC content in NIP hydrogels. According to the elemental analysis data, higher degree of incorporation of MAC into the NIP hydrogels was realized by using sulfur stoichiometry which was 2.36% for NIP and 1.83% for MIP–Cd(II) hydrogels. In the case of pNIPA, very low adsorption capacity was observed compared to MAC-incorporated hydrogels. Lower adsorption capacity of the pNIPA hydrogels indicated that poorer binding ability of the pNIPA (-N atom of the amide group in NIPA) than those of MAC groups toward Cd(II) ions. From the findings that we obtained in this study, the new polymeric imprinted hydrogels presented in this communication are promising for the adsorption of Cd(II) ions from aqueous media.

6.3.2.4 Temperature-dependent adsorption

The amount of cadmium ions adsorbed as a function of temperature, and NIP and MIP–Cd(II) hydrogels at pH 5 are presented in Figure 6.10. The initial Cd(II) concentration was 10 ppm and the adsorption time was 24 h. As shown in Figure 6.10, both NIP and MIP–Cd(II) hydrogels show high adsorption amount for Cd(II) ions at low temperature, 22 °C (below LCST), and the adsorbed amount of Cd(II) ions per unit mass of the hydrogels decreases with increasing the temperature (above LCST). The results indicated that the hydrophilic-to-hydrophobic transition of polymer

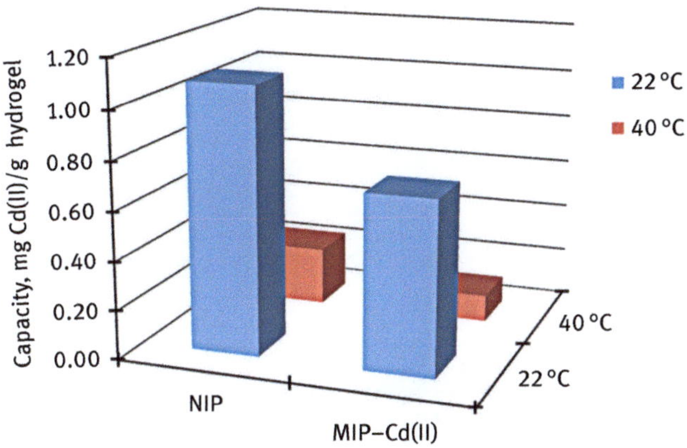

Figure 6.10: Effect of temperature on the adsorption amount for Cd(II) ions of the NIP and MIP–Cd(II) hydrogels.

networks triggered by the temperature increase affected the adsorption of Cd(II) ions. Thus, the higher adsorption capabilities of both NIP and MIP–Cd(II) hydrogels toward Cd(II) ions below LCST can be attributed to the fact that the adsorption of the hydrogels mainly depend on the complexation of Cd(II) ions with MAC groups. The "swollen–shrunken" configuration change of NIP-based hydrogel networks triggered by environmental temperature could influence the formation of MAC–Cd(II) complexes. At temperatures lower than the LCST, the copolymer networks stretch, which makes it easier for the inner MAC groups to capture the Cd(II) ions, so that the hydrogels exhibit higher adsorption capacity. On the other hand, at temperatures above LCST, the NIP networks shrink and the inner MAC groups are close to each other. As a result, the electrostatic repulsions among the ions affect the formation of stable MAC–Cd(II) complexes inside hydrogel, which leads to smaller adsorbed amount of Cd(II). From a practical point of view, the results demonstrate that the available adsorption sites decrease with changes in hydrogel dimension produced by temperature variation.

6.3.3 Selectivity experiments

In this group of experiments, competitive adsorption of Cd(II), Pb(II), and Cu(II) ions and Cd(II), Cr(III), and Fe(III) from their separate solutions or mixed ion solutions was investigated. For this purpose, competitive adsorption studies were performed in two ways. First, adsorption studies were done with separate solutions of Cd(II), Pb(II), Cu(II), Cr(III), and Fe(III) ions with each solution containing 10 ppm of each metal ions, at pH: 5.5 with a stirring rate of 350 rpm at 22 °C for 24 h (Figure 6.11). Second, the mixture of Cd(II), Pb(II), and Cu(II) solutions contain

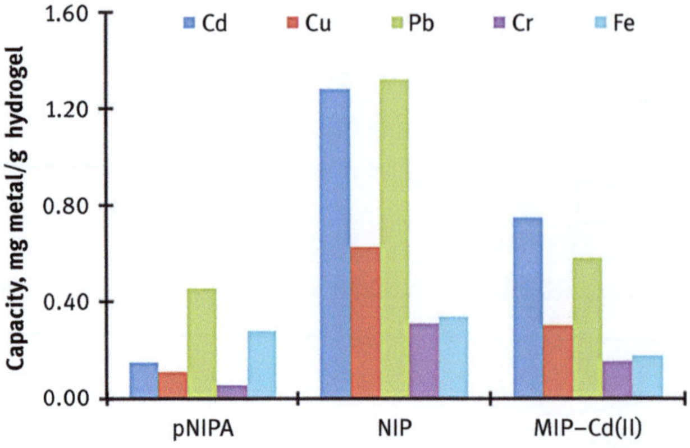

Figure 6.11: Metal ions adsorption from singular aqueous solutions.

10 ppm of each metal ion, and Cd(II), Cr(III), and Fe(III) solutions also contain 10 ppm of each metal ions; other adsorption conditions were kept the same as in the case of separate solutions of these metal ions (Figure 6.12). The measurements were carried out with the MIP–Cd(II) and NIP hydrogels. Figure 6.12 shows the results of the selective adsorption of Cd(II) ions in competitive solutions. For comparison, noncompetitive adsorption results of each metal ion are given in Figure 6.11.

As can be seen from the data in Figure 6.11, under noncompetitive conditions, the metal ion adsorption amount for the MIP–Cd(II) hydrogel increased in the order of Cd (II) > Pb(II) > Cu(II) > Fe(III) > Cr(III). On the other hand, in the case of the NIP hydrogel the adsorption amount of Pb(II) ions (1.32 mg/g) was a little large compared with Cd(II) ions (1.28 mg/g). These results support the ion-recognition behavior of the MIP–Cd(II) hydrogel. According to the Pearson acid–base rules, complexation behavior of ligands and cations in terms of electron pair donating Lewis bases and electron pair accepting Lewis acids, metal ions are termed hard, soft, and borderline. Hardness of metal ions (Lewis acids) will determine their preference to binding. Softer ions (e.g., Cd(II), low positive charge relative to large size, very polarizable) are expected to bind sulfur and nitrogen donor atoms of the ligand on the polymer, whereas hard metals (e.g., Cr(III), high charge to radius ratio, not very polarizable) coordinate to carboxylate groups, and borderline ions (e.g., Cu(II) and Pb(II)) would bind to any of the ligands according to conditions that may change the hardness of the ligand. On the other hand, under competitive conditions, less amount of all these ions were adsorbed to the MIP–Cd(II) and NIP hydrogels because of competitions of these ions. However, the selective adsorption of the Cd(II) ions on the MIP–Cd(II) hydrogel was observed in the same manner as in Figure 6.12. This can be concluded that NIP and MIP–Cd(II) hydrogels show the following metal ion affinity in the order of Cd(II) > Pb(II) > Cu(II) > Cr(III) > Fe(III). From these results, it can be said that the hydrogel adsorbent imprinted with Cd(II) ions indicates the selectivity for the Cd(II) ion as expected.

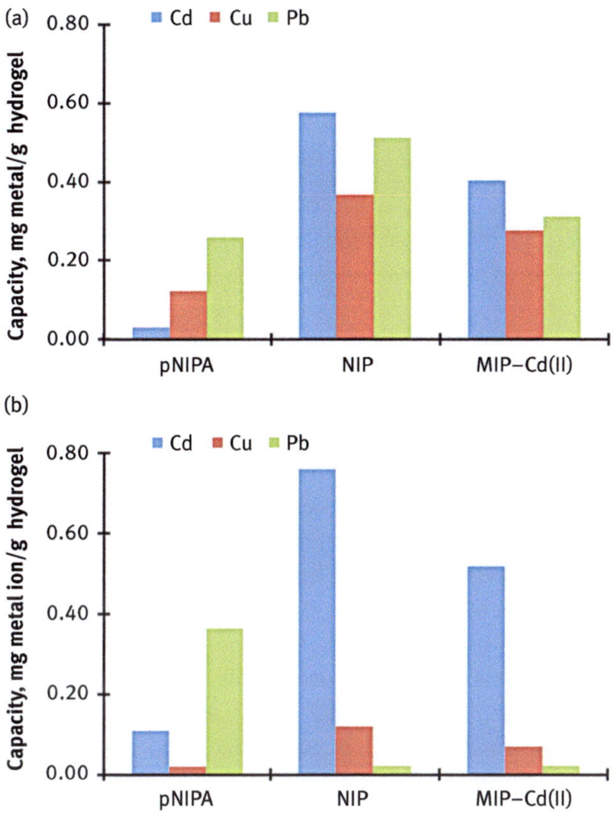

Figure 6.12: Competitive heavy metal adsorption from aqueous solutions: (a) triple metal ion solution containing Cd(II), Cu(II), and Pb(II); and (b) triple metal ion solution containing Cd(II), Cr(III), and Fe(III).

6.3.4 Desorption and reusability

With the aim of developing polymeric hydrogels sensitive to external (temperature) stimuli and able to reversibly adsorb and release Cd(II) ions, we tried to take out the Cd(II) ions from the MIP–Cd(II) hydrogel by squeezing the gels above LCST. First, the NIP and MIP–Cd(II) hydrogels were incubated in 10 ppm Cd(II) solution at 22 °C for 12 h. After the adsorption was completed, adsorption medium was placed in a thermostatic water bath at 50 °C for 24 h. The results of desorption of Cd(II) ions, by squeezing the NIP and MIP–Cd(II) hydrogels at 50 °C, were given in Table 6.1. According to our expectation, we could shrink the swollen hydrogels despite the presence of MAC–Cd(II) complexes by raising the temperature to 50 °C. Unfortunately, both the NIP and the MIP–Cd(II) hydrogels released water when shrinking, but did not release Cd(II) ions at mentionable amount. The imprinted

Table 6.1: Reusability of the imprinted and nonimprinted hydrogels; adsorption–desorption of Cd (II) ions from the hydrogels.

Adsorbent	Cd(II) ions adsorbed, mg/g			Desorption ratio, %		
	First	Second	Third	First	Second	Third
MIP–Cd(II)	0.727	0.691	0.709	98.9	98.6	99.2
NIP	1.204	1.31	1.116	99.7	97.9	97.2

hydrogel released just 6.02% of the adsorbed Cd(II) ions, which means the squeezed hydrogel still keeps 93.98% of the bound Cd(II) ions in the polymeric hydrogel network, and the NIP hydrogel released 13.49% of the adsorbed Cd(II) ions. The results can be concluded that there is a frustration for adsorbing monomers to come into proximity, which comes from the cross-links and the polymer connection, so that the affinity is recovered upon shrinking. Consequently, the Cd(II) ions adsorbed on the hydrogels were completely desorbed from the hydrogels by means of acid treatment. About 3 h of interaction time with 0.1 M HNO_3 was enough for complete desorption and the amounts of the stripping Cd(II) ions were almost the same with the adsorbed amounts in the experimental error limits (Table 6.1).

In order to investigate the reusability of the MIP–Cd(II) and NIP hydrogels, adsorption–desorption cycle was repeated three times using the same sorbent. The data are presented in Table 6.1, which indicated that resorption capacity of the hydrogels for Cd(II) ions did not change significantly during repeated adsorption–desorption operations. Thus, it is concluded that the NIP and MIP–Cd(II) hydrogels can be used many times without decreasing their adsorption capacities significantly.

Alvarez-Lorenzo et al. prepared copolymer gels of NIPA and methacrylic (MAA) monomers to reversibly adsorb and desorb divalent ions, and Cd(II) was chosen as a target divalent atom. To enhance the affinity to calcium, they applied imprinting technique using Cd(II) and Pb(II) as templates. They reported that (a) the affinity depends on the degree of gel swelling or shrinkage that can be switched on and off by temperature (b) in the shrunken state, affinity depends approximately linearly on the MAA concentration in the MIP–Cd(II) gels, whereas in the NIP gels it is proportional to the square of MAA concentration (c); the MIP–Cd(II) hydrogels adsorb more than the NIP gels when MAA concentration is less than that of permanent cross-linkers [29]. Yamashita et al. examined the preparation of IPN-type stimuli-responsive gel consisting of pNIPAm and pNaAAC for heavy-metal-ion adsorption/desorption [30].

They pronounced that the swelling IPN gel below the LCST quickly adsorbed the Cu(II) ions but did not release Cu(II) ions at all. In another study, Ju et al. showed the adsorption/desorption of lead(II) ions from aqueous solutions by using the p(NIPAM-co-BCAm) hydrogels. They observed complete desorption of Pb(II) ions above the LCST [11].

6.3.5 Determination of Cd(II) ions in a certified sample

Using a certified water sample (NWTMDA-52.3 fortified water, obtained from LGC Standards), the recovery of the Cd(II) ions from the nonimprinted and the imprinted hydrogels was verified. The content of the certified sample is given in Table 6.2. The adsorption studies were carried out in 50 mL aliquots of the certified sample with a stirring rate of 300 rpm and adjusted the pH to 5.5 at 22 °C for 12 h. For the desorption of the Cd(II) ions, 50 mL of 0.1 M HNO_3 was used. The certified value and the obtained results after desorption with HNO_3 are presented in Table 6.3. As can be seen from the data in Tables 6.2 and 6.3, both NIP and MIP–Cd(II) hydrogels exhibited a satisfactory recovery values in spite of the presence of other metal ions.

Table 6.2: The content of the certified water sample, NWTMDA-52.3.

Element	Concentration, µg/L	Element	Concentration, µg/L
Ag	20.8	Li	14.9
Al	310	Mn	198
As	25.3	Mo	208
Ba	148	Ni	274
Be	17.6	Pb	358
Bi	12.5	Sb	16.5
Cd	190.9	Se	21.8
Co	136	Sn	19.9
Cr	165	Sr	286
Cu	189	Ti	120
Fe	413	Tl	18.4

Table 6.3: Results for determination of Cd(II) ions in certified water sample.

Hydrogel sample	Certified value for Cd, µg/L	Determined value for Cd, µg/L	Recovery, %
MIP–Cd(II)	190.9	164.6	86.2
NIP	190.0	178.3	93.3

6.4 Conclusions

In this study, we have proposed ion-imprinted smart hydrogels having the ability to respond temperature variation. Although we could not achieve the efficient releasing of template ions via temperature variation, which was contrary to general expectation, the main results reported here showed specific recognition of template ions by imprinted hydrogels. The imprinting process also improved the structural and stimuli-responsive properties of synthesized hydrogels. Therefore, the proposed hydrogels could be classified as smart polymers based on their stimuli responsibility and ion recognition abilities.

References

[1] Hoffman, A.S., New antibody purification procedure using thermally responsive poly(NIPA)-dextran derivative conjugate, Journal of Chromatography B, 2001, 761, 247–254.

[2] Lee, W.F., and Lin, Y.H. Swelling behavior and drug release of NIPAAm/PEGMEA copolymeric hydrogels with different crosslinkers, Journal of Materials Science, 2006, 41, 7333–7340.

[3] Turmanova, S., Vassilev, K., and Boneva, S., Preparation, structure and properties of metal-copolymer complexes of poly-4-vinylpyridine radiation-grafted onto polymer films, Reactive Functional Polymers, 2008, 68, 759–767.

[4] Suzuki, D., and Kawaguchi, H., Hybrid microgels with reversibly changeable multiple brilliant color, Langmuir, 2006, 22, 3818–3822.

[5] Vassilev, K., Turmanova, S., Dimitrova, M., and Bonev, S. Poly(propylene imine) dendrimer complexes as catalysts for oxidation of alkenes, European Polymer Journal, 2009, 45, 2269–2278.

[6] Yeghiazarian, L., Mahajan, S., Montemagno, C., Cohen, C., and Wiesner, U. Directed motion and cargo transport through propagation of polymer-gel volume phase transitions, Advanced Materials, 2005, 17, 1869–1873.

[7] Hoffman, A.S. Intelligent polymers in medicine and biotechnology, Macromolecular Symposia, 1995, 98, 645–664.

[8] Hu, L., Chu, L.Y., Yang, M., Wang, H.D., and Niu, C.H. Preparation and characterization of novel cationic pH-responsive poly(N, N-dimethylamino ethyl methacrylate) microgels, Journal of Colloid and Interface Science, 2007, 311, 110–117.

[9] Kanazawa, R., Mori, K., Tokuyama, H., and Sakohara, S. Preparation of thermosensitive microgel adsorbent for quick adsorption of heavy metal ions by a temperature change, Journal of Chemical Engineering of Japan, 2004, 37, 804–807.

[10] Tokuyama, H., Kanazawa, R., and Sakohara, S. Equilibrium and kinetics for temperature swing adsorption of a target metal on molecular imprinted thermosensitive gel adsorbents, Separation and Purification Technology, 2005, 44, 152–159.

[11] Ju, X.J., Zhang, S.B., Zhou, M.Y., Xie, R., Yang, L., and Chu, L.Y. Novel heavy metal adsorption material: ion-recognition P(NIPAM-co-BCAm) hydrogels for removal of lead(II) ions, Journal of Hazardous Materials, 2009, 167, 114–118.

[12] Kanazawa, R., Mori, K., Tokuyama, H., and Sakohara, S. Preparation of thermosensitive microgel adsorbent for quick adsorption of heavy metal ions by a temperature change, Journal of Chemical Engineering of Japan, 2004, 37, 804–807.
[13] Denizli, A., Garipcan, B., Karabakan, A., Say, R., Emir, S., and Patir, S. Metal-complexing ligand methacryloylamidocycteine containing polymer beads for Cd(II) removal, Separation and Purification Technology, 2003, 30, 3–10.
[14] Bae, Y.H., Okano, T., and Kim, S.W. Temperature dependence of swelling of crosslinked poly (N,N-alkyl substituted acrylamide) in water, Journal of Polymer Science: Polymer Physics, 1990, 28, 923–936.
[15] Arora, H., Malik, R., Yeghiazarian, L., Cohen, C., and Wiesner, U. Earthworm inspired locomotive motion from fast swelling hybrid hydrogels, Journal of Polymer Science, Polymer Chemistry, 2009, 47, 5027–5033.
[16] Wang, B., Xu, X.D., Wang, Z.C., Cheng, S.X., Zhang, X.Z., and Zhu, R. Synthesis and properties of pH and temperature sensitive P(NIPAAm-co-DMAEMA) hydrogels, Colloids and Surfaces B, 2008, 64, 34–41.
[17] Zhang, J., and Peppas, N.A. Synthesis and characterization of pH- and temperature sensitive poly(methacrylic acid)/poly(N-isopropylacrylamide) interpenetrating polymeric networks, Macromolecules, 2000, 33, 102–109.
[18] Yu, Y., Chang, X., Ning, H., and Zhang, S. Synthesis and characterization od thermoresponsive hydrogels cross-linked with chitosan, Central European Journal of Chemistry, 2008, 6, 107–113.
[19] Chaterji, S., Kwon, I.K., and Park, K., Smart polymeric gels: redefining the limits of biomedical devices, Progress in Polymer Science, 2007, 32, 1083–1122.
[20] Zhang, X.Z., and Chu, C.C. Preparation of thermosensitive PNIPAAm hydrogels with superfast response, Chemical Communications, 2004, 2004, 350–351.
[21] Zhang, X.Z., and Chu, C.C. Influence of polyelectrolyte on the thermosensitive property of PNIPAAm-based copolymer hydrogels, Journal of Materials Science: Materials in Medicine, 2007, 18, 1771–1779.
[22] Hirashima, Y., and Suzuki, A. Formation and destruction of hydrogen bonds in gels and in aqueous solutions of N-isopropylacrylamide and sodium acrylate observed by ATR-FTIR spectroscopy, Journal of Colloid and Interface Science, 2007, 312, 14–20.
[23] Iyer, G., Viranga Tillekeratne, L.M., Coleman, M.R., and Nadarajah, A. Equilibrium swelling behavior of thermally responsive metal affinity hydrogels, Part II: Solution effects, Polymer, 2008, 49, 3744–3750.
[24] Zhang, X.Z., and Chu, C.C., Fabrication and characterization of microgel-impregnated thermosensitive PNIPAAm hydrogels, Polymer, 2005, 46, 9664–9673.
[25] Andac, M., Say, R., and Denizli, A. Molecular recognition based cadmium removal from human plasma, Journal of Chromatography B, 2004, 811, 119–126.
[26] Andac, M., Ozyapı, E., Senel, S., Say, R., and Denizli, A., Ion-selective imprinted beads for aluminum removal from aqueous solutions, Industrial & Engineering Chemistry Research, 2006, 45, 1780–1786.
[27] Andac, M., Mirel, S., Senel, S., Say, R., Ersoz, A., and Denizli, A. Ion-imprinted beads for molecular recognition based mercury removal from human serum, International Journal of Biological Macromolecules, 2007, 40, 159–166.
[28] Demircelik, A.H., Andac, M., Andac, C.A., Say, R., and Denizli, A. Molecular Recognition-Based Detoxification of Aluminum in Human Plasma, Journal of Biomaterials Science, 2009, 20, 1235–1258.

[29] Alvarez-Lorenzo, C., Guney, O., Oya, T., Sakai, Y., Kobayashi, M., Enoki., T., Takeoka, Y., Ishibashi, T., Kuroda, K., Tanaka, K., Wang, G., and Grosberg, A.Y. Reversible adsorption of calcium ions by imprinted temperature sensitive gels, Journal of Chemical Physics, 2001, 114, 2812–2816.

[30] Yamashita, K., Nishimura, T., and Nango, M. Preparation of IPN-type stimuli responsive heavy-metal-ion adsorbent gel, Polymers for Advanced Technologies, 2003, 14, 189–194.

Nityananda Agasti

7 Ag and Au nanoparticles for detection of heavy metals in water

7.1 Introduction

Heavy metals can be defined as metals with atomic weights between 63.5 and 200.6 g mol^{-1} and specific gravity greater than 5 g cm^{-3} [1]. Contamination of water by heavy metals is one of the most serious environmental concerns, which poses grave threat to living organisms. In the era of development to cater the versatile need of human population, industrial activities increase manifold. Due to industrial activities heavy metals enter into water bodies. Heavy metals such as Hg, As, Pb, and Cd are highly toxic and carcinogenic even at a trace level [2–3]. Heavy metals enter human body through various sources and disrupt cellular functions leading to toxicity and get excreted through liver, kidney, or spleen. Toxicity of heavy metals is due to their bond formation with thiol groups of proteins and enzymes, leading to inhibition of enzymes. This can be explained on the basis of Pearson acid–base concept, hard soft acid–base theory (HSAB), which describes that soft acids react faster with soft bases and hard acids react faster with hard bases. Heavy metals like Hg, Pb, and As being soft acids form stable complexes with soft base like S. Heavy metals also cause toxicity through oxidative stress and impaired antioxidant metabolism, thereby affecting human health. They are nonbiodegradable and pose a severe threat to human health. It is therefore essential to detect heavy metals in drinking water. Conventional methods used for detection of heavy metal ions include graphite furnace atomic absorption spectroscopy, cold vapor generation, and an ion chromatograph pretreatment system with inductively coupled plasma emission spectroscopy. Methods like atomic absorption spectrometry, fluorescent sensors, colorimetric sensors, electrochemical sensors, X-ray absorption fine structure spectroscopy, and ultrasensitive dynamic light scattering assays are used to detect heavy metals.

Although conventional methods were used for detection of heavy metals in water, but tedious sample preparation, sophisticated instrumentation, and lack of portability demand devices that can provide quick and simple methods for detection of heavy metals in water. Nanomaterials have shown great potential for quick, reliable, and simple methods for detection of heavy metals. Nanomaterials as sensors have provided new opportunities for detection of heavy metals with high sensitivity, selectivity, and detection limit up to very low concentration. Integration of

Nityananda Agasti, Department of Chemistry, Deen Dayal Upadhyaya College, University of Delhi, New Delhi, India

https://doi.org/10.1515/9783110650600-007

nanomaterials for fabrication of sensing devices makes the detection process more portable and handy. Metal nanoparticles (NPs), semiconductor NPs, carbon nanotubes, and so on are used as sensors for detection of heavy metals.

This chapter covers noble metal NPs like Ag and Au as sensors for detection of heavy metals like Pb, Hg, and As in water. Based on their sensing mechanism, sensors can be summarized, including optical, electrochemical, and field-effect transistor (FET) sensors. The optical sensors include fluorescent, colorimetric, surface plasmon resonance (SPR), and surface-enhanced Raman scattering sensors. In this chapter, the use of Au and Ag NPs as electrochemical and colorimetric sensor for detection of heavy metal in water is discussed.

7.2 Electrochemical detection

Heavy metals like Pb, Hg, and As can be detected by electrochemical sensors with Au and Ag NPs. Electrochemical sensors help in analyzing the chemical reactions by electrical means. Electrochemical sensing involves several techniques, including voltammetry, amperometry, potentiometry, impedemetry, and conductometry. In particular, the anodic stripping voltammetry (ASV) method is readily amendable for determination of heavy metals. ASV analysis typically involves two steps: (i) electrochemical deposition or accumulation of heavy metals at a constant potential onto the electrode surface and (ii) stripping or dissolution of the deposited analyte from the electrode surface. Although bulk electrodes have been used in ASV analysis of heavy metals, the lowest detection limit and the sensitivity of bulk electrodes cannot meet the need for detection of trace heavy metals. Metal NP-modified electrodes are found to be a suitable alternative to bulk electrodes for detection of heavy metals.

The addition of metallic NPs such as Au and Ag helps in increasing the ability to trigger specific reactions and electron transfer rate between analytes and electrode and also helps in avoiding undesirable products. Electrochemical sensors coupled with metal NPs like Au and Ag offer distinctive advantages such as high sensitivity and fast real-time detection. Gold (Au NPs) and silver nanoparticles (Ag NPs) immobilized on the surface of an electrode can be used as electrochemical sensors for heavy metal detection. NPs can be immobilized on the surface of an electrode by dipping a thoroughly cleaned electrode into the colloidal solution of Au and Ag NPs stabilized by capping agents such as citrate [4], amino acid [5], thiol [6], and so on. Electrocatalytic reaction, due to oxidation and reduction of heavy metal ion catalyzed by Au and Ag NPs, can take place on the surface of the electrode and thus makes the detection of analyte (heavy metal ion) possible. The designing of a sensor with best-suited architecture helps in selective identification of heavy metal ions in water.

7.3 Colorimetric detection

Au and Ag NPs act as a promising colorimetric sensor because of their high sensitivity due to very high extinction coefficient in the visible region owing to the distinctive property of SPR. SPR arises from collective oscillation of free electrons in the conduction band of noble metal in resonance with the incident electromagnetic radiation [7]. Au NPs exhibit SPR at a wavelength range of 450 to 700 nm and Ag NPs exhibit SPR at a wavelength range of 400 to 530 nm. These SPR bands are highly sensitive toward interparticle distance, size of the NPs, and local environment around NPs. Change in local environment around NP changes the SPR band of NPs and thus shows color tunable behavior of Ag and Au NPs. Since the SPR band of metal NPs is dependent on interparticle distance, so aggregation of Au and Ag NPs causes change in color and SPR band. Aggregation of NPs can be induced by interaction between chromophores and analytes or metal ions. Based on this principle, Au and Ag NPs act as colorimetric sensor. Variation in the concentration of analyte also changes the color of NPs. Au and Ag NPs functionalized with suitable ionophores can selectively bind with analyte or metal ion, resulting in aggregation of NPs. Aggregation of NPs leads to color change and thus makes them an efficient sensor. Ionophores are molecules that have functional groups such as –SH, $-NH_2$, which can bind to the surface of Au and Ag NPs and a coordinating site for interaction with analyte metal ion.

7.4 Gold nanoparticles (Au NPs) for detection of Hg^{2+}, Pb^{2+}, and As^{3+} in water

7.4.1 Electrochemical detection

7.4.1.1 Detection of Hg^{2+}

Au NPs adsorbed on solid electrodes were used for electrochemical detection of heavy metals in water sample. Ratner and Mandler [8] reported Au NPs deposited on glassy carbon (GC) and indium tin oxide (ITO) electrodes as an effective sensing device for detection of Hg^{2+} in water. Au NPs were produced by electrochemical reduction of $AuCl_4^-$ and stabilization by citrate. Au NPs were adsorbed on GC and ITO electrodes by dipping thoroughly cleaned GC and ITO electrodes into Au NP solution for a given time. Linear sweep voltammetry (LSV) technique has been used for detection of Hg^{2+} in water. LSV of GC and ITO electrodes modified with Au NPs in an electrolyte of 0.1 M KCl was observed both in the presence and absence of Hg^{2+}. After adding Hg^{2+} into electrolyte, LSV of GC and ITO electrodes shows deposition of Hg on the electrode surface and gives stripping peaks corresponding to the reduction of Hg^{2+}. Stripping peaks and deposition of Hg when compared both in Au NP modified electrode and bare electrode suggest that Hg sensing is significantly improved by adsorbing Au NPs on electrode. Therefore, Au NPs play an important role in deposition and stripping of Hg. Au NPs serve as a nucleation site for electrodeposition of Hg. By using Au NPs modified GC and ITO electrode, Hg^{2+} in water samples can be detected up to a limit of 1 µm.L^{-1}.

A carbon ionic liquid nanocomposite electrode modified with glutathione capped Au NPs has been used for electrochemical detection of Hg^{2+} in water by Safavia and Farjamib [9]. Glutathione-capped Au NPs were prepared by citrate reduction method. Carbon ionic liquid electrode was prepared by using graphite and octylpyridinum hexafluorophosphate ($OPy+PF_6^-$) with a ratio of 50:50(w/w). Nanocomposite electrode was prepared by mixing graphite powder, ionic liquid, and glutathione-capped Au NPs (50:40:10, wt%) [10]. Using square wave voltammetry with glutathione-capped Au NPs composite carbon ionic liquid electrode as the working electrode in aqueous solution of Hg^{2+}, a stripping peak is observed in the voltammogram corresponding to the presence of mercury. However, the stripping peak is not observed in voltammogram of aqueous solution of Hg^{2+} with bare carbon ionic liquid electrode not modified with glutathione-capped Au NPs. Thus, the role of Au NPs in detection of Hg^{2+} is ascertained. The lowest detection limit of Hg^{2+} in water through this composite electrode is found to be 2.3 nM.

7.4.1.2 Detection of Pb^{2+}

Au NPs immobilized on a surface can be used as a potential sensor for electrochemical detection of Pb^{2+} in water. Electrochemical detection of Pb^{2+} in water by using Au NPs assembled FET device has been reported by Chen et al. [11]. A FET device assembled with reduced graphene oxide (rGO) and glutathione functionalized Au NPs can be used as sensor for Pb^{2+} in aqueous solution. rGO as the semiconducting channel material is utilized in the FET device through a self-assembly method.

As shown in Scheme 7.1, the device is made up of gold electrodes assembled on silicon substrate. The GO sheets were selectively deposited onto the Au electrodes by a self-assembly method. To anchor GO on Au electrode surface, a layer of α-ethyl-tryptamine (AET) is deposited on Au electrode surface by immersing Au electrode in a solution of AET. Then by emerging Au electrode in GO solution, GO sheets self-assembled onto the surface of Au electrode. Thermal reduction was conducted by heating the electrode and Au NPs were deposited on rGO by sputtering technique. Finally Au NPs were functionalized with glutathione by immersing the device in a solution of glutathione (10 mM). The FET device was washed with deionized water and dried before use. In the presence of Pb^{2+} the electrical characteristics of FET device change and thus allow the detection of Pb^{2+}. The FET characteristic changes upon the introduction of the Pb^{2+} solutions of concentrations ranging from 10 nM to 10 mM, and only takes a few seconds to

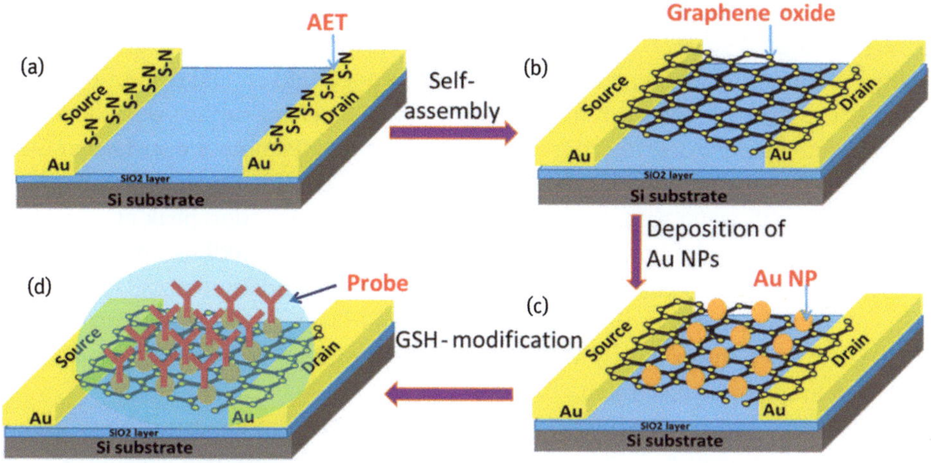

Scheme 7.1: Illustration of the rGO/GSH-Au NP hybrid sensor fabrication process. (a) A layer of AET coating on the bare interdigitated electrode surface. (b) Self-assembly of GO monolayer sheets on the AET-modified electrodes, which is followed by the thermal reduction of GO to rGO. (c) The assembly of Au NPs onto the rGO film. (d) GSH modification of Au NPs on the rGO sheet surface to form specific recognition groups to detect Pb^{2+}. Reprinted with permission from G. Zhou, J. Chang, S. Cui, H. Pu, Z. Wen, J. Chen, *ACS Appl. Mater. Interfaces* 2014, **6**, 19235, Copyright © 2014, American Chemical Society [11].

respond. Thus, rGO/GSH-Au acts as a selective and sensitive sensor for detection of Pb^{2+} in aqueous solution.

Au NPs assembled on the surface of glassy carbon electrode (GCE) to form an Au NP/GC electrode can be used for detection of Pb^{2+} in water. Differential pulse ASV technique can be used for detection of Pb^{2+} by using Au NP/GC electrode. Cai et al. have developed Au NP/GC electrode for detection of Pb^{2+} [12]. Colloidal Au NPs were deposited onto the clean and dry surface of GCE by electrodeposition method. The cleaned GCE and platinum wire electrode were used as the anode and cathode under a potential of 1.0 V for 90 minutes immersed in as-prepared colloidal Au NPs solution. After electrodeposition, the Au NP/GC electrode was rinsed with deionized water and dried. Differential pulse ASV was used with Au NP/GC electrode to detect Pb^{2+} in an aqueous solution up to a detection limit of 3×10^{-7}M. The voltammetric signals of bare GCE and Au NPs deposited GCE were observed. Higher anodic and cathodic peak current with Au NPs-modified GCE than bare GCE suggests the role of Au NPs for detection of Pb^{2+} in water.

7.4.1.3 Detection of As^{3+}

Electrochemical technique has been an area of interest for detection of As^{3+} in water due to high sensitivity and quick detection up to a very low concentration.

Au NPs immobilized on GCE have been used as a sensor for detection of As^{3+} by ASV technique [13]. ASV was conducted with electrode modified with Au NPs in water containing As^{3+}. Clean and dry GCE was dipped into colloidal solution of citrate-capped Au NPs. GCE with Au NPs adsorbed on its clean surface was allowed to dry at room temperature. In order to firmly adhere Au NPs onto the surface of GCE, chitosan (CT) solution was transferred to Au NPs-loaded GCE. Cyclic voltammetry of Au NP/GCE with water sample containing As^{3+} gives a reduction peak at ~ −0.05 V corresponding to reduction of As^{3+} to As^0. On the reverse anodic scan, voltammogram gives oxidation peak at +0.18 V corresponding to oxidation of As^0 to As^{3+}. An increase in concentration of As^{3+} from 1 to 5 ppm in water sample leads to a linear increase in both the peaks. Lower detection limit of As^{3+} in water by Au NP/GCE has been found to be 0.025 ppb.

Chen et al. [14] reported Au NPs crystal violet (CRV) film deposited on GCE as a sensor for detection of As^{3+} in drinking water. CRV is an organic dye. Electrochemical deposition of Au NPs and CRV was done on the GCE using cyclic voltammetry. Modified GCE with nano-Au-CRV film was then washed with double-distilled water before its use as As^{3+} sensor. Electrochemical detection of As^{3+} in water is made possible by cyclic voltammetry using nano-Au-CRV-modified GCE. Appearance of anodic oxidation peaks in voltammogram suggests detection of As^{3+} in water. In addition, the anodic oxidation peak in voltammogram is found to be linearly dependent on the concentration of As^{3+}. Null response toward As^{3+} in voltammogram with bare GCE and CRV-modified GCE shows that Au NPs are responsible for detection As^{3+} in water. The lowest electrochemical detection limit of As^{3+} with nano-Au-CRV/GCE is found to be 0.2 μM.

Electrocatalytic oxidation of As^{3+} to As^{5+} in the presence of Au NPs can be a strategy for detection of As^{3+} in water. Nanocomposite film was prepared by layer-by-layer assembly of citrate-capped Au NPs with cationic polyelectrolytes [15]. The catalytic activity of the Au nanocomposite has been exploited for As^{3+} detection by electrocatalytic oxidation of As^{3+} to As^{5+}. A gold disk electrode cleaned and dried was dipped in a solution of 20 mM sodium 3-mercapto-1-propanesulfonate (MPS) prepared in 16 mM H_2SO_4 for half an hour to form a self-assembled monolayer of MPS on the gold surface. The negatively charged MPS-modified gold electrode (Au-MPS) was then immersed in a solution of positively charged polyelectrolyte PDDA (poly(diallyldimethylammonium chloride)) for 15 min. The electrode with a monolayer of PDDA was then soaked in a dispersion of citrate-capped Au NPs for 90 min to obtain a layer of Au NPs. Repetition of this cycle results in a nanocomposite Au-MPS-(PDDA–Au $NPs)_5$, where five layers of PDDA–Au NPs are deposited on a gold foil electrode. Differential pulse voltammetry of As^{3+} solution with a nanocomposite of PDDA–Au NPs multilayer assembly as the working electrode, a Pt wire as the counter electrode, and Ag-AgCl (3 M KCl) as the reference electrode was carried out. The electrocatalytic oxidation of As^{3+} takes place on the layer-by-layer assemblies. The appearance of oxidation peaks in voltammogram indicates oxidation of As^{3+} to

As^{5+}, thereby detecting the presence of As^{3+} in water sample. An increase in the number of Au NPs in nanocomposite increases the catalytic activity. A comparison between nanocomposite of PDDA–Au NPs multilayer assembly and bare Au electrode for electrocatalytic oxidation of As^{3+} suggests better performance of nanocomposite than that of bare Au electrode. The detection limit of As^{3+} in water is found to be 4.336 ppb, which is well below the WHO guideline of 100 ppb in drinking water.

7.4.2 Colorimetric detection

7.4.2.1 Detection of Hg^{2+}

Au NPs exhibit SPR absorption in the visible wavelength range. This SPR band is sensitive toward local environment of NPs and suffers a shift upon interaction of NPs with external species. This shift in SPR band with a change in local environment makes Au NPs a potential colorimetric sensor for detection of mercuric ions in an aqueous medium. Y. Guo et al. [16] reported papain-coated Au NPs as a sensor for detection of Hg^{2+} in water sample. Papain-coated Au NPs change their color from red to blue upon interaction with Hg^{2+}. This color change takes place because of the aggregation of NPs due to a binding between Hg^{2+} and papain. This binding is due to high affinity of Hg^{2+} toward S atom in papain. This aggregation leads to a substantial shift in SPR band to a longer wavelength, thus making the detection of Hg^{2+} in water possible. Wu et al. [17] reported 8-hydroxy-quinoline (8-HQ)-modified Au NPs stabilized by polyvinylpyrrolidone (PVP) as a highly selective and sensitive colorimetric detector of Hg^{2+}. Au NPs tend to aggregate quickly in the presence of Hg^{2+}, leading to a color change of the NP solution from wine red to gray, along with the change in both, the SPR absorption and intensity. This aggregation of NPs takes place because of Hg^{2+} binding to quinoline N and carbonyl oxygen atoms. It has also been observed that colorimetric response of unmodified Au NPs toward Hg^{2+} ions has been enhanced by the addition of a molecule that promotes aggregation of NPs. Denizli et al. [18] observed that the addition of lysine molecule promotes aggregation of Au NPs, leading to a highly sensitive detection of Hg^{2+} in water sample. Hg^{2+} was added to citrate-stabilized Au NPs and then lysine was added to NP solution. The addition of lysine induced aggregation of Au NPs because the amino group of lysine binds to Hg^{2+}. This aggregation of Au NPs leads to a change in color and shift in SPR absorption. The mechanism of sensing Hg^{2+} by Au NPs in the presence of lysine can be illustrated as in Scheme 7.2.

Hg^{2+} in water sample can also be detected by using colorimetric Au NPs on a paper-based analytical device. Chen et al. [19] reported DNA-functionalized Au NPs immobilized on paper for detection of Hg^{2+} in water. When aqueous solution of Hg^{2+} is added into Au NPs, thymine base pair of DNA binds to Hg^{2+} ion, which leads to aggregation of Au NPs. Thus color and SPR of Au NPs gets changed. The change in

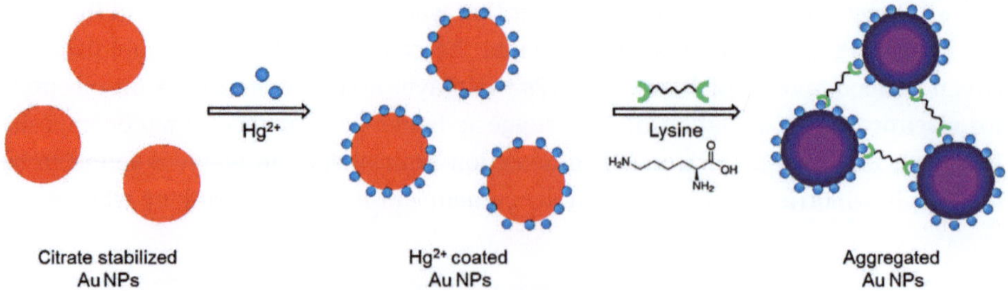

Scheme 7.2: Proposed Hg^{2+} sensing mechanism of the colorimetric assay. Reprinted with permission from Gulsu Sener, Lokman Uzun and Adil Denizli, *Analytical Chemistry*, 2014, **86**, 514, Copyright © 2014, American Chemical Society [18].

color of Au NPs allows easy detection of Hg^{2+} in an aqueous medium. Using this paper-based device, the lower detection limit of Hg^{2+} can be observed to be 50 nM.

SPR of Au NPs varies with their size and shape, giving rise to change in color of Au NP solution. So morphology-dependent color change of nano-Au can be utilized for detection of analyte, for example, Hg^{2+} in water. Han et al. [20] reported detection of Hg^{2+} in water by change in shape of nano-Au from rod shape to spherical shape denoted by color change. Hexadecyltrimethylammonium bromide-coated gold nanorods (AuNRs) in the presence of ascorbic acid (AA) in an aqueous solution of Hg^{2+} changes their morphology to spherical Au NPs, thus facilitating detection of Hg^{2+} in water. Au NRs exhibit two characteristic absorption bands at long and short wavelengths related to longitudinal and transverse modes, respectively. When aqueous solution of Hg^{2+} is added to AuNRs in AA solution, the absorption band corresponding to longitudinal mode disappears, giving rise to a single SPR band. This single band corresponds to spherical Au NPs. This shift in SPR can be detected by spectrophotometer and by naked eye due to color change. This color change is the basis for Au NPs to be used as a sensor. Thus, AuNRs detect Hg^{2+} in an aqueous solution in the presence of AA with a low detection limit 1 μM by naked eye and 30 nM by spectrophotometer.

7.4.2.2 Detection of Pb^{2+}

Analyte-induced aggregation of Au NPs, leading to a change in color and shift in SPR absorption, can be employed for detection of Pb^{2+} in water. Tseng et al. [21] used gallic acid-capped Au NPs for colorimetric detection of Pb^{2+} in water sample. The addition of Pb^{2+} into gallic acid-capped Au NPs leads to aggregation of Au NPs. This aggregation is due to coordination of Pb^{2+} with phenolic hydroxyl group of gallic acid. Au NPs assembled on a membrane can be used as a sensor for colorimetric detection of Pb^{2+} in water sample. A paper-based colorimetric sensor has been developed by Lee and Huang [22] for rapid detection of Pb^{2+}. Paper-based membrane contains Au NPs-modified nitrocellulose. The nitrocellulose membrane (NCM) has been used to trap Au NPs modified with bovine serum albumin (BSA) through hydrophobic interaction between BSA and NCM.

Au NPs were prepared through citrate reduction of $HAuCl_4$ [23]. Aqueous solution of 1 mM $HAuCl_4$ (250 mL) was brought to a vigorous boil by stirring in a round-bottom flask fitted with a reflux condenser; 38.8 mM trisodium citrate (25 mL) was added rapidly to the solution and the solution was heated for another 15 min, turning from pale yellow to deep red. The solution was cooled to room temperature by stirring continuously. Au NPs prepared were modified with BSA by putting an aliquot of BSA solution to Au NPs solution and equilibrated for 30 minutes at 25 °C. Then to trap BSA-modified Au NPs on NCM, a piece of NCM was dipped into the solution of BSA-modified Au NPs. After air drying at room temperature a nanocomposite was prepared containing BSA-modified Au NPs assembled on NCM. By putting an aliquot of aqueous solution of Pb^{2+} on the membrane BSA-Au-NCM in the presence of $Na_2S_2O_3$ and 2-mercaptoethanol, the color of the membrane changes, leading to naked eye detection of Pb^{2+} in water. Here the color change takes place because of leaching of Au NPs by Pb^{2+}. Leaching of Au NPs was accelerated by $S_2O_3^{2-}$ and 2-mercaptoethanol. $S_2O_3^{2-}$ ion forms Au $(S_2O_3)_2^{3-}$ complex on the Au NP surfaces. The addition of Pb^{2+} ions and 2-mercaptoethanol induced the deposition of Pb^{2+} on the surfaces of the Au NPs. The deposition of Pb^{2+} accelerated the dissolution of the Au NPs, forming Au^+-2-mercapto ethanol complexes in the solution [24]. As a result, the SPR absorbance of the BSA-Au NPs at 520 nm decreases, leading to a change in color of Au NPs and facilitates the detection of Pb^{2+}. The mechanism of sensing of Pb^{2+} can be schematically represented in Scheme 7.3.

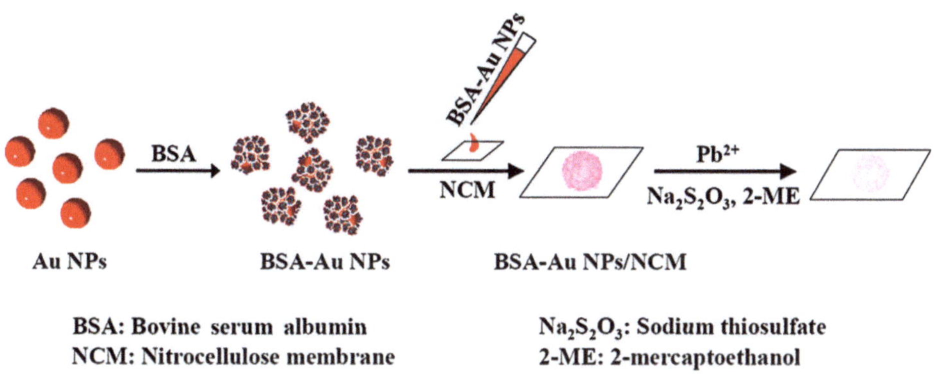

Scheme 7.3: Schematic representation of the preparation of BSA-Au NPs/NCM for sensing lead ions (Pb^{2+}) based on accelerated leaching rate of BSA-Au NPs by sodium thiosulfate and 2-mercaptiethanol. Reprinted with permission from Y. F. Lee and C. C. Huang, *Appl. Mater. Interfaces,* 2011, **3**, 2747, Copyright © 2011, American Chemical Society [22].

7.4.2.3 Detection of As^{3+}

Since SPR band of Au NPs are dependent on interparticle distance, the variation in colorimetric signals of Au NPs will take place based on their transformation from dispersed to aggregated form. Because of this colorimetric signal variation, Au NPs can be used as a potential material for sensing applications. Surfactant-induced aggregation of Au NPs can be used to sense As^{3+} in aqueous solution. Wu et al. [25] developed a biosensor for detection of As^{3+} based on the aggregation of Au NPs that is controlled by the special interactions among the arsenic binding aptamer, hexadecyltrimethylammonium bromide (HTAB) and As^{3+}. In the presence of As^{3+}, the aptamer forms an aptamer–As^{3+}complex, so that the subsequent HTAB addition aggregates the Au NPs, leading to remarkable variations of absorbance. However in the absence of As^{3+} aptamers remain free and can form some supramolecule with HTAB. Since HTAB is bound to aptamer, it cannot aggregate Au NPs and thus absorbance of dispersed Au NPs will be different. Therefore, aptamer and HTAB-induced aggregation and nonaggregation of Au NPs make them a potential candidate for As^{3+} detection in water. Using this HTAB-induced aggregation of Au NPs, the lowest detection limit of As^{3+} in water has been found to be 40 ppb for naked eye and 0.6 ppb for spectrophotometer. Sensing of As^{3+} in aqueous solution by Au NPs is illustrated in Scheme 7.4.

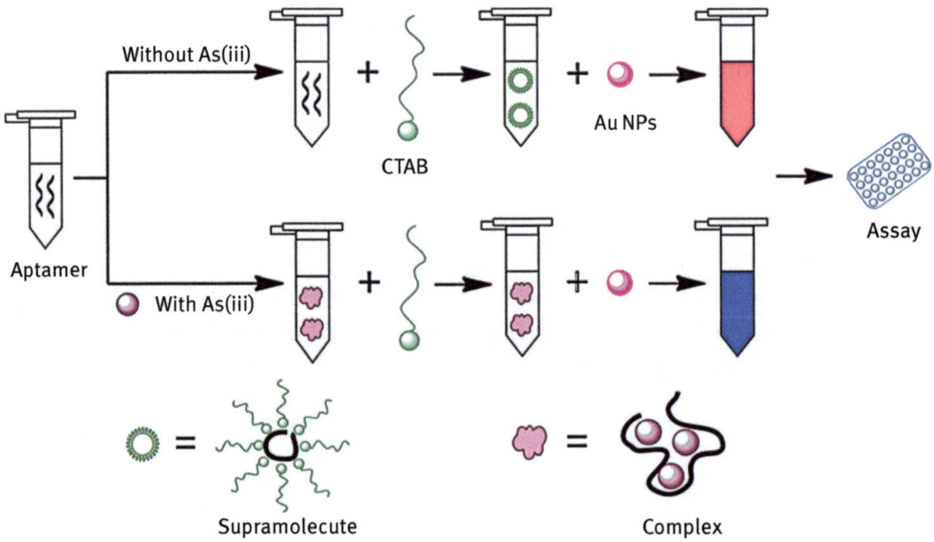

Scheme 7.4: Illustration of the biosensor for As(III) detection based on the surfactant-induced aggregation of Au NPs. Reprinted with permission from Yuangen Wu, Le Liu, Shenshan Zhan, Faze Wang and Pei Zhou, *Analyst*, 2012, **137**, 4171, Copyright © 2012, Royal Society of Chemistry [24].

7.5 Silver nanoparticles (Ag NPs) for detection of Hg^{2+}, Pb^{2+}, and As^{3+} in water

Ag NPs can be used for detection of heavy metals Hg^{2+}, Pb^{2+}, and As^{3+} in water sample. Electrochemical and colorimetric techniques are among important methods for detection of heavy metals in aqueous medium discussed here.

7.5.1 Electrochemical detection

7.5.1.1 Detection of Pb^{2+}

Ag NPs-modified electrodes are used as an electrochemical sensor for detection of Pb^{2+} in water. Ag NPs immobilized on GCE is used for detection of Pb^{2+} up to picomolar concentration range [26]. On a clean and dry GCE, PVP-stabilized Ag NPs are deposited using chronoemperometry. Ag NP/GC electrode is used for detection of Pb^{2+} by square wave ASV. The cyclic voltammogram of water sample containing Pb^{2+} with Ag NPs-modified GCE as a working electrode shows stripping peaks corresponding to lead ion. However cyclic voltammogram of bare GCE does not show stripping peak corresponding to lead ion. So the deposition of Ag NPs on GCE enhances the sensitivity of electrode for Pb^{2+} detection. With Ag NP/GC electrode as a sensor, the lowest detection limit of Pb^{2+} in water is obtained as 10 pM.

7.5.1.2 Detection of As^{3+}

Ag NPs can be used for detection of As^{3+} in water by voltammetry. Ag NPs immobilized on the surface of an electrode can be used in voltammetry to detect As^{3+} in water. Shahi et al. [27] reported the use of a modified GCE comprising of Ag NPs adsorbed on CT for detection of As^{3+} by differential pulse anodic stripping voltammetry. CT, a suitable biopolymer having good adhesion to electrode surfaces, exhibits its high water permeability, nontoxicity, and biocompatibility. Hydrophilic surface of CT due to the presence of reactive amino and hydroxyl functional groups makes it useful for developing electrochemical sensors and biosensors. Ag NPs embedded in CT matrix was prepared by reduction of $AgNO_3$ by hydrazine hydrate, followed by stabilization through CT matrix [27]. About 10 ml of 30 mmol L^{-1} $AgNO_3$ was added to 10 ml of 5 mg/ml CT stock solution under stirring (30 min) at room temperature. Then 1 µl of hydrazine hydrate (35%, v/v) was added to the resulting mixture. The mixture was kept under constant stirring at 60 C for 3 h. Material was obtained by centrifugation and washed a couple of times with water. The purified sample was dispersed in water. By putting drops of suspension of CT-Ag NPs on the surface of a clean and dry GCE, Ag NP/CT-modified GCE was prepared. Differential pulse ASV of water sample containing As^{3+} with Ag NP/CT-modified GCE was conducted. Stripping peak corresponding to reduction of As^{3+} to As0 was observed in voltammogram. However, voltammogram of water sample containing As^{3+} shows no stripping peak with unmodified GCE in the absence of Ag NP. This shows the role of Ag NPs for detection of As^{3+} in water.

Srivastava and coworkers [28] reported β-cyclodextrin-stabilized Ag NP and graphene oxide nanocomposite (Ag NPs-GO) as a highly sensitive and selective electrochemical sensor for detection of As^{3+} in the presence of other elements in water sample. The nanocomposite Ag NPs-GO was immobilized on the surface of GCE by the drop deposition method to construct β-cyclodextrin-stabilized Ag NPs–GO/GCE. Cyclic voltammetry and ASV were used for the electrochemical sensing property of β-cyclodextrin-stabilized Ag NPs–GO/GCE for detection of As^{3+} in water. The nanostructured electrode exhibits high sensitivity and selectivity toward As^{3+} in water with a lower detection limit of 0.24 nM.

7.5.2 Colorimetric detection

7.5.2.1 Detection of Hg^{2+}

Ag NPs are used as optical probes for detection of Hg^{2+} in water. Chakraborty et al. [29] reported cysteamine-capped Ag NPs as a colorimetric sensor for Hg^{2+} ions in water with a lower detection limit 0.273 nM. When colloidal solution of cysteamine-capped Ag NPs is added to the water sample containing Hg^{2+}, surface property of Ag NPs changes, leading to a change in color of Ag NPs. On treatment with Hg^{2+}, thiol group of cysteamine binds with Hg^{2+}. Then a redox reaction involving Ag^0 and Hg^{2+} would lead to the formation of Ag–Hg nanoalloy (eq. 7.1). As a result the surface plasmon property of Ag Nps changes, leading to a change in the color of the solution and thus making Ag NPs an efficient colorimetric probe for Hg^{2+}. The mechanism of Hg^{2+} sensing is depicted in Scheme 7.5

$$Ag_n + Hg^{2+} \rightarrow Ag_{n-2}Hg + 2Ag^+ \qquad (7.1)$$

Since SPR of Ag NPs is dependent on interparticle distance, the aggregation and antiaggregation of Ag NP change the color of Ag NP solution. Hg^{2+}-induced antiaggregation of Ag NPs can be utilized for detection of Hg^{2+} in water. Duan et al. [30] reported a colorimetric sensor for Hg^{2+} in water based on antiaggregation of Ag NPs. Colloidal solution of citrate-capped Ag NPs suffers color change from yellow to brown by the addition of 6-thioguanine due to aggregation of Ag NPs. Thiol group of 6-thioguanine binds to Ag and thus causes aggregation of Ag NPs. When Ag NPs containing 6-thioguanine is added to water sample containing Hg^{2+}, the color of Ag NPs solution changes from brown to yellow, indicating antiaggregation of

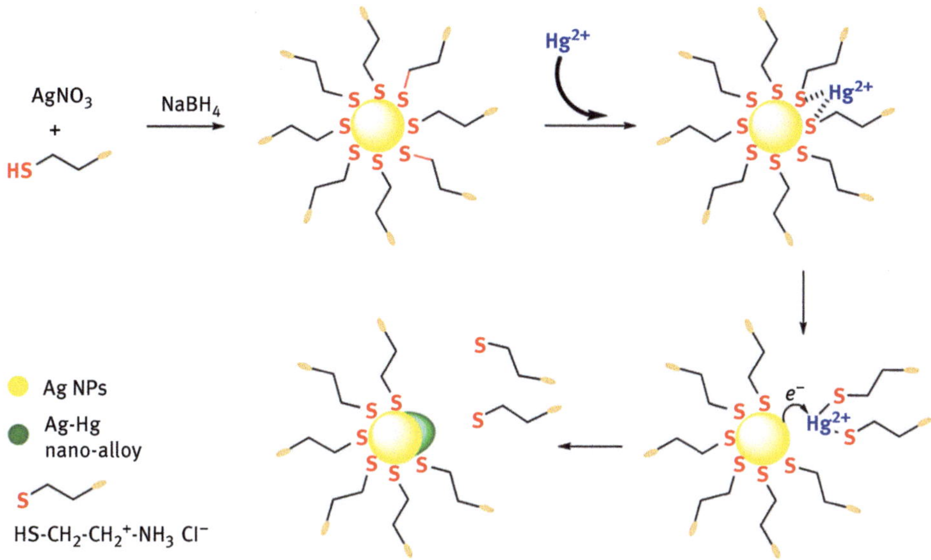

Scheme 7.5: Representation of sensing mechanism of Hg^{2+} ions. Reprinted with permission from Yudhajit Bhattacharjee and Amarnath Chakraborty, *Sustainable Chem. Eng.* 2014, **2**, 2149, Copyright © 2014, American Chemical Society [28].

aggregated Ag NPs. Hg^{2+}-induced antiaggregation is due to a greater affinity of thiol group toward Hg^{2+} than Ag. This can be explained on the basis of HSAB principle. Sulfur being a soft base will bind to a soft acid Hg^{2+} more strongly. Here 6-thioguanine and Hg^{2+} act as an aggregation reagent and antiaggregation reagent, respectively. The changes in color of Ag NP solution induced by the competitive interactions of 6-thioguanine with Ag NPs and Hg^{2+} make the colorimetric detection of Hg^{2+} in water possible. By this method, the detection limit of Hg^{2+} is found to be 4 nM. Agasti et al. [31] have used UV-Visible absorption spectroscopic method to detect Hg^{2+} ions in water using Ag nanoparticles. Change in SPR band of Ag nanoparticles with concentration of Hg^{2+} has been useful for making the detection process simple and fast.

7.5.2.2 Detection of Pb^{2+}

Ag NPs are used as a colorimetric sensor for Pb^{2+} because of their distance-dependent optical properties. Interaction of Pb^{2+} with surface of Ag NPs changes the SPR band of Ag NPs and thus enables Ag NPs to act as a sensor. Since sensitivity of Ag NPs is directly related to its surface, the surface functionalization of Ag NPs improves their selectivity, consequently widening their applications as probe to detect Pb^{+2}. Therefore, it is important to modify the surface of Ag NPs with ligands that can bind to analyte or Pb^{2+} and make Ag NPs aggregated. Aggregation of Ag NPs leads to a change in color and absorption spectrum of Ag NPs, thus giving a colorimetric response. Wu et al. [32] reported iminodiacetic acid (IDA)-capped Ag NPs as a colorimetric probe for Pb^{2+}. Since IDA has a strong binding ability with Pb^{2+}, it induces aggregation of Ag NPs when IDA-capped Ag NPs were added to water containing Pb^{2+}. Because of aggregation, the color of IDA-capped Ag NPs solution gets changed. The change in color of NP solution in the presence of Pb^{2+} in water can be detected.

Aggregation-induced shift in SPR band of Ag NPs has been utilized to detect Pb^{2+} in water. Ag NPs stabilized by biocompatible molecule like gallic acid have been used as a sensor for detection of Pb^{2+} in aqueous medium [33]. Gallic acid acts as both reducing agent and stabilizing agent. Gallic acid contains both carboxylic acid group and phenolic group. Carboxylic acid group binds to Ag NPs and thus stabilizes Ag NPs from agglomeration. When Pb^{2+} is added to gallic acid-capped Ag NPs, phenolic hydroxyl group of Ag NPs surface bound gallic acid binds to Pb^{2+}. As a result Ag NPs become closer to each other. The close proximity of NPs induces coupling of their plasmon oscillation, resulting in a bathochromic shift in the absorption band. In the presence of Pb^{2+}, a visual color change of the Ag NPs solution from yellow to red is observed, thus making the detection of Pb^{2+} in water easy.

7.5.2.3 Detection of As^{3+}

Like Hg^{2+} and Pb^{2+} SPR of Ag NPs is also sensitive toward As^{3+}. Interaction of As^{3+} with surface-modified Ag NPs changes the color and SPR of NPs and thus acts as a colorimetric probe for As^{3+} in water. Divsar et al. [34] reported aptamer-conjugated silver nanoparticles (Apt-Ag NPs) for colorimetric detection of Pb^{2+}. Aptamer-capped Ag NPs was synthesized. The addition of water sample containing As^{3+} ions into Apt-Ag NPs solution causes an interaction of As^{3+} with Apt-Ag NPs to form As^{3+}–Apt-Ag NPs. As a result a remarkable shift in the SPR of Ag NPs takes place, which thus enables the determination of As^{3+} with high selectivity and sensitivity. So the colorimetric detection is a simple and efficient method to detect As^{3+} in water with a lower limit of 6 µg L^{-1}. The mechanism of colorimetric sensing of As^{3+} is illustrated in Scheme 7.6.

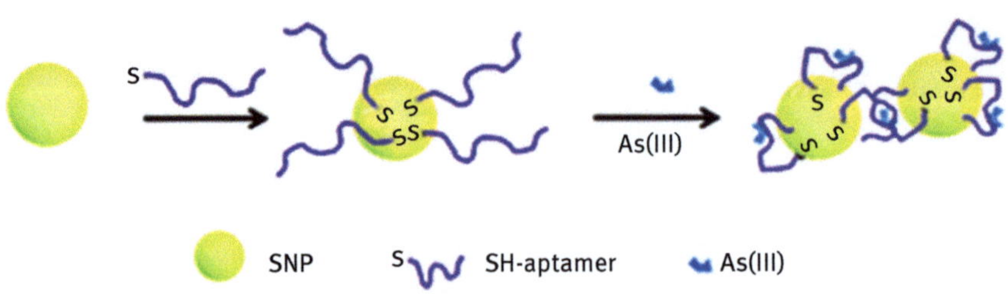

Scheme 7.6: Representation of sensing mechanism of As^{3+} ions. Reprinted with permission from F. Divsar, K. Habibzadeh, S. Shariati and M. Shahriarinour, *Anal. Methods.* 2015, **7**, 4568, Copyright © 2015, Royal Society of Chemistry [33].

References

[1] Turdean, G.L. International Journal of Electrochemical, 2011, 1.
[2] Duffus, J.H. Pure and Applied Chemistry, 2002, 74, 793.
[3] Aragay, G., Pons, J., and Merkoçi, A. Chemical Reviews, 2011, 111, 3433.
[4] Flores, C.Y., Diaz, C., Rubert, A., Benítez, G.A., Moreno, M.S., Fernández Lorenzo de Mele, M.A., and Vericat, C. Journal of Colloid and Interface Science, 2010, 350, 402.
[5] Agasti, N., Singh, V.K., and Kaushik, N.K., Materials Research Bulletin, 2015, 64, 17.
[6] Kumar, S., Bolan, M.D., and Bigioni, T.P. Journal of the American Chemical Society, 2010, 132, 13141.
[7] Mayer, K.M., and Hafner, J.H. Chemical Reviews, 2011, 111, 3828.
[8] Ratner, N., and Mandler, D. Analytical Chemistry, 2015,
[9] Safavia, A., and Farjamib, E. Analytica Chimica Acta, 2011, 688, 43.
[10] Maleki, N., Safavi, A., and Tajabadi, F. Analytical Chemistry, 2006, 78, 3820.
[11] Zhou, G., Chang, J., Cui, S., Pu, H., Wen, Z., and Chen, J. ACS Applied Materials & Interfaces, 2014, 6, 19235.
[12] Xiaoxia, X., Duan, G., Li, Y., Liu, G., Wang, J., Zhang, H., and Cai, W. ACS Applied Materials & interfaces, 2014, 6, 65.
[13] Lan, Y., Luo, H., Ren, X., Wang, Y., and Wang, L. Analytical Letters, 2012, 45, 1184.
[14] Rajkumar, M., Thiagarajan, S., and Chen, S.-M. International Journal of Electrochemical Science, 2011, 6, 3164.
[15] Ottakam Thotiyl, M.M., Hajra Basit, J.A., Sánchez, C.G., Coche-Guerente, L., Pascal Dumy, S., Sampath, P.L., and Moutet, J.-C. Journal of Colloid and Interface Science, 2012, 383, 130.
[16] Guoa, Y., Wang, Z., Qub, W., Shao, H., and Jiang, X. Biosensors and Bioelectronics, 2011, 26, 4064.
[17] Gao, Y., Li, X., Li, Y., Tianhua, L., Zhao, Y., and Aiguo, W. Chemical Communications, 2014, 50, 6447.
[18] Sener, G., Uzun, L., and Denizli, A. Analytical Chemistry, 2014, 86, 514.
[19] Yen, Y.-C., Wang, C.-W., Chang, H.-T., and Chen, C.-F. Analytical Chemistry, 2014, 86, 6843.
[20] Jin, L.-H., and Han, C.-S. Sensors and Actuators B, 2014, 195, 239.
[21] Huanga, K.-W., Yua, C.-J., and Tseng, W.-L., Biosensors and Bioelectronics, 2010, 25, 984.
[22] Lee, Y.F., and Huang, C.C. ACS Applied Materials & Interfaces, 2011, 3, 2747.
[23] Jana, N.R., Gearheart, L., and Murphy, C.J. Journal of Physical Chemistry B, 2001, 105, 4065
[24] Chen, -Y.-Y., Chang, H.-T., Shiang, Y.-C., Hung, Y.-L., Chiang, C.-K., and Huang, -C.-C. Analytical Chemistry, 2009, 81, 9433.
[25] Yuangen, W., Liu, L., Zhan, S., Wang, F., and Zhou, P. Analyst, 2012, 137, 4171.
[26] Sivasubramanian, R., and Sangaranarayanan, M.V. Talanta, 2011, 85, 2142.
[27] Prakash, S., Chakrabarty, T., Singh, A.K., and Shahi, V.K. Electrochimica Acta, 2012, 72, 157.
[28] Dar, R.A., Khare, N.G., Cole, D.P., Karna, S.P., and Srivastava, A.K., RSC Advances 2014, 4, 14432.
[29] Bhattacharjee, Y., and Chakraborty, A. ACS Sustainable Chemistry & Engineering, 2014, 2, 2149.
[30] Duan, J., Yin, H., Wei, R., and Wang, W. Biosens Bioelectron, 2014, 15(57), 139.
[31] Agasti, N., and Kaushik, N.K. Advanced Materials Letters, 2018, 9(1), 53.
[32] Li Qi, Y., and Shang, F.W. Microchimica Acta, 2012, 178, 221.
[33] Yoosaf, K., Ipe, B.I., Suresh, C.H., and Thomas, K.G. Journal of Physical Chemistry C, 2007, 111, 12839.
[34] Divsar, F., Habibzadeh, K., Shariati, S., and Shahriarinour, M. Analytical Methods, 2015, 7, 45–68.

Index

https://doi.org/10.1515/9783110650600-008

Bei Fragen zur Produktsicherheit wenden Sie sich bitte an:
If you have any questions regarding product safety,
please contact:

Walter de Gruyter GmbH
Genthiner Straße 13
10785 Berlin
productsafety@degruyterbrill.com